EAST AFRICAN WILDLIFE

A VISITOR'S GUIDE

T0244540

PHILIP BRIGGS &
ARIADNE VAN ZANDBERGEN

www.bradtguides.com

Bradt Guides Ltd, UK
The Globe Pequot Press Inc, USA

Bradt GUIDES

TRAVEL TAKEN SERIOUSLY

Third edition published November 2024
First published October 2007
Bradt Travel Guides Ltd
31a High Street, Chesham, Buckinghamshire, HP5 1BW, England
www.bradtguides.com
Print edition published in the USA by The Globe Pequot Press Inc,
PO Box 480, Guilford, Connecticut 06437-0480

Text copyright © Philip Briggs, 2024
Maps copyright © Bradt Travel Guides Ltd, 2024; includes map data © OpenStreetMap contributors
Photographs copyright © Individual photographers, 2024 (see below)
Project Manager: Susannah Lord
Cover research: Pepi Bluck, Perfect Picture

ISBN: 9781804692226

British Library Cataloguing in Publication Data
A catalogue record for this book is available from the British Library

Photographs Ariadne Van Zandbergen, www.africaimagelibrary.com (AVZ);
Africa Imagery: Andrew Woodburn/Liquipi (AW/L/AI); Roger de la Harpe (RH/AI);
Dreamstime.com: Henkbogaard (H/D), Jonathan Caisley (JC/D); Stephen Spawls (SSp);
SuperStock (SS); Wikimedia Commons: Lord Mountbatten (LM/WC)
Front cover Lioness in the grass (AVZ)
Back cover, clockwise from top left Plains zebra (*Equus quagga*), Akagera National Park, Rwanda;
African elephant (*Loxodonta africana africana*); leopard (*Panthera pardus*) (all AVZ)
Title page, left to right White-fronted bee-eater (*Merops bullockoides*); Maasai giraffe (*Giraffa
camelopardalis tippelskirchi*), at Lake Ndutu at sunset, Serengeti National Park, Tanzania (both AVZ)

Maps Malcolm Barnes, updated by David McCutcheon FBCart.S. FRGS

Typeset by Ian Spick, Bradt Guides
Production managed by Gutenberg Press Ltd; printed in Malta
Digital conversion by www.dataworks.co.in

Paper used for this product comes from sustainably managed forests,
and recycled and controlled sources.

UPDATES WEBSITE

For the latest on-the-ground travel news, trip reports and factual updates,
go to w bradtguides.com/updates. If you have any comments, queries,
insights, news or other feedback, you can contact us on 01753 893444 or
e info@bradtguides.com. We will forward emails to the authors who may
post updates on the Bradt website. Alternatively, you can add a review of the
book to Amazon, or share your adventures with us on Facebook, Twitter or
Instagram (@BradtGuides).

CONTENTS

AUTHOR AND MAJOR CONTRIBUTORS

Philip Briggs is a travel and environmental writer specialising in Africa. Born in Britain and raised in South Africa, he started travelling in East Africa in 1986, and his first book *Guide to South Africa* was published by Bradt Travel Guides in 1991. Since then, Philip has divided his time between exploring and writing about the highways and byways of Africa. He is the author or co-author of more than a dozen Bradt travel guides, covering Uganda, Tanzania, Rwanda, Ghana, Ethiopia, Malawi, Kenya, South Africa, Somaliland, Gambia, Suriname, Sri Lanka and Mozambique. He has also contributed to numerous other books and magazines.

Ariadne Van Zandbergen, who took most of the photographs for this book and contributed to the research, is a freelance photographer and tour guide. Born and raised in Belgium, she travelled through Africa from Morocco to South Africa in 1994/5 and is now resident in Wilderness, South Africa. She has visited 25 African countries and her photographs have appeared in numerous books, magazines, newspapers, maps, periodicals and pamphlets. In addition to co-photographing the sumptuous coffee-table book *Africa: Continent of Contrasts* (Struik/New Holland) with Martin Harvey, she runs her own online photo library (w africaimagelibrary.com).

Mike Unwin, who provided the text on which the Invertebrates chapter of this book is based, is the author of the Bradt companion book *Southern African Wildlife: A Visitor's Guide*, and writes widely on African wildlife and travel.

ACKNOWLEDGEMENTS

This book is the product of four decades of on-off travel through East Africa, much of which was undertaken to research other guidebooks and articles, with the support of innumerable local safari operators, lodges and other institutions that were acknowledged at the time and to which I extend my ongoing gratitude. But, as befits a project that has gestated over so many years, I'd like to take this opportunity to thank a few people whose support has been more enduring in its nature: my parents Roger and Kay Briggs for countless lifts to and from airports; my wife Ariadne Van Zandbergen for dragging me to ever more remote corners of Africa in the name of photography; Hilary and the rest of the Bradt team; and editor Mike Unwin for allowing me to base the Invertebrate section in this book on the corresponding text in his companion guide.

EAST AFRICA

AFRICA

SOUTH SUDAN

ETHIOPIA

3187m
Lokichogio
Akobo
Dawa
Lake Turkana
Mandera
Moyale
Marsabit
Wajir

Gulu
Mt Moroto 3084m
Turkwe
Kerio
Pakwach
Lira
2445m
Lake Albert
Albert Nile
Lake Kyoga
Mt Elgon 4321m
3206m
Ewaso Nagiri

UGANDA
Kafu
Victoria Nile
Eldoret
KENYA
Garissa
Tana

KAMPALA
Jinja
Tororo
Kisumu
Nakuru
Mt Kenya 5199m
3098m
3999m

Kasese
Masaka
Entebbe
Lake Victoria
Mara
NAIROBI

Lake Edward
Mbarara
Akagera
Lake Natron 3648m
Kilimanjaro 5895m
Namanga
Garsen
Lamu

Kabale
RWANDA
Lake Kivu
KIGALI
Arusha
Athi
Moshi
Voi
Mombasa

Cyangugu
BURUNDI
Mwanza
Pangani
Pemba

DEMOCRATIC REPUBLIC OF CONGO

Kigoma
Ugalla
Tabora
Tanga
Zanzibar

Lake Tanganyika
2373m
DODOMA
Morogoro
Dar es Salaam
Mafia

Mpanda
TANZANIA
Rungwa
2646m
INDIAN

Lake Rukwa
Njombe
Great Ruaha
Rufiji
OCEAN

Mbeya
Kilombero
Luwegu
Matandu
Lindi

2961m
ZAMBIA
MALAWI
Songea
Masasi
Ruvuma

1893m
Lake Nyasa

MOZAMBIQUE

SOMALIA

KEY to maps on pages 249–70
Road or track
Trail
Airport (international)
Summit (height in metres)
Area of Interest (e.g. Kilombero Valley)
CA = Conservation Area
ES = Elephant Sanctuary
FR = Forest Reserve
GR = Game Reserve
GS = Game Sanctuary
MNP = Marine National Park
MNR = Marine National Reserve
NP = National Park
NR = National Reserve
WC = Wildlife Conservancy
WR = Wildlife Reserve
WS = Wildlife Sanctuary

0   250km
0   250 miles

N
Bradt!

KEY
Capital
Town
Airport (international)
Mountain peak
Land over 3000m

INTRODUCTION

East Africa exists on the grandest imaginable scale. Geographically, it is dominated by the vast tectonic scar known as the Great Rift Valley, the Earth's only non-aquatic natural feature that is visible from the moon. Scenically, East Africa's breathtaking landmarks range from snow-capped Kilimanjaro to the lava-spewing Virungas; from the oceanic expanses of lakes Victoria and Tanganyika to an endless succession of idyllic Indian Ocean beaches; from the parched plains of northern Kenya to the steamy rainforests of western Uganda. Culturally, its diversity embraces relict hunter-gatherers such as the Hadza and Batwa, red-robed traditional pastoralists such as the Maasai and Samburu, the Arab-tinged Swahili of Zanzibar and Lamu, and the centralised medieval kingdoms of Uganda – not to mention the cosmopolitan modernity of cities such as Nairobi, Kigali, Kampala and Dar es Salaam.

Elephant bulls in the Ngorongoro Crater. (AVZ)

Above all, however, East Africa is renowned for its peerless concentrations of wildlife, epitomised by the annual wildebeest migration across the Serengeti-Mara plains and the million-strong flamingo flocks that form a pink swathe around lakes Nakuru and Bogoria. And it's not only the Serengeti: there's also the immense Nyerere and Ruaha national parks protecting a combined area of more than 50,00km^2 in southern Tanzania, the buffalo-studded plains of Katavi, the semi-arid badlands of Samburu-Buffalo Springs, and the yawning hippos and armour-plated crocs that swim in the Nile below Murchison Falls – not to mention the primate-rich rainforests that line the Congolese border, with their habituated mountain gorillas and chimpanzees.

In recent years, the global media have routinely portrayed East Africa as being in the throes of a conservation crisis. In a sense, this is true. Population growth, habitat loss and hunting have taken a heavy toll on the region's wildlife over the past century, and deforestation remains rife outside protected areas. But East Africa also remains one of the last places on Earth where the immense mammal herds that once roamed our entire planet still exist in something like their prehistoric numbers, protected within a network of 100 odd national parks and other conservation areas that are managed but not sublimated by the most numerous and destructive large mammal species of all. Conservation crisis? Perhaps, but that's largely because East Africa, compared with practically anywhere else you might care to mention, still has so many wild places and wild creatures left to conserve!

Of course, one need look no further than the precipitous numerical decline over the past century of such flagship species as African wild dog, black rhino and mountain gorilla to recognise the vulnerability of East Africa's natural abundance – especially when it is required to meet the ever-growing demands of a human population that increased tenfold during the 20th century and continues to double every two decades. In this context, tourism, far from being the ecological threat that some have alleged, might well prove integral to the survival of East Africa's wildlife. Not only does tourism generate direct revenue in the form of entrance fees, but it also provides conservation-related employment in areas where job opportunities are limited, and is a strong deterrent to poachers, who cannot easily go about their clandestine business in areas regularly visited by tourists.

East Africa's wildlife sanctuaries are a varied bunch. Some, like Nyerere or Tsavo, are as vast as a small European country, others little bigger than a suburban park. Many are managed by the relevant state authority with or without the backing of international NGOs, but an increasing number operate as low-key ecotourism sites managed by local communities, or leased from them for a substantial fee. Some comprise endless vistas of parched savannah, while others protect isolated montane, forest or wetland habitats. There's little doubt, however, that this superb network of sanctuaries is the repository of the world's most bountiful and exciting wildlife.

The red-chested sunbird (*Cinnyris erythrocerus*) is a typically colourful member of a group of nectar-eaters reminiscent of the New World hummingbirds. (AVZ)

The world's tallest land mammal, the giraffe (*Giraffa camelopardalis*) is regularly encountered in national parks such as Tsavo East in southern Kenya. (AVZ)

ABOUT THIS BOOK

Several travel guides are available to help you get around East Africa (which for present purposes comprises Tanzania, Kenya, Uganda and Rwanda), while a selection of excellent field guides can assist you in identifying every bird, mammal, reptile and even amphibian you might see in the wild. This book aims to bridge the gap between these more specialised books by providing a compact and handy 'one-stop shop' reference to the region's wildlife. It is aimed at visitors whose curiosity about what they see on safari might not easily be sated by local guides, but also doesn't quite extend to being weighed down with a veritable library of hefty and costly tomes.

The core of the book consists of four chapters each dedicated to a principal taxonomic group, namely mammals, birds, reptiles and amphibians, and invertebrates. These do not attempt to replicate the function of a field guide by focusing solely on identification, but instead attempt to describe the more common species or genera, placing them in an evolutionary context and highlighting their more interesting behavioural traits. Coverage focuses on 'typical' savannah wildlife associated with the popular safari reserves of Tanzania and Kenya, but also looks in some detail at the plethora of more elusive creatures that inhabit the region's other habitats. To keep things manageable and retain the focus on terrestrial safaris, freshwater fish are omitted on the basis that they are not easily observed in their natural habitat, and marine wildlife on the basis that it warrants a whole book in its own right.

Other shorter chapters place the wildlife in context: *The Lie of the Land* (page 7) provides an overview of the geography and climate of East Africa, while *Habitats* (page 17) describes the main wildlife-viewing environments. *On Safari* (page 241) details some of the more popular activities and how to make the most of them, and includes the section *Where to go* (page 248), which takes a quick tour of all the important wildlife reserves in each of the four countries covered. Finally, a short *Further Reading* (page 274) section points you towards more specialised travel and field guides, and other books to help you plan and enjoy your trip.

Lions, though generally terrestrial, regularly climb trees in a few specific locations, notably Tanzania's Lake Manyara and Serengeti national parks, and in Ishasha in Uganda. (AVZ)

THE LIE
OF THE LAND

Maasai giraffe with young in
Mkomazi National Park. (AVZ)

GEOGRAPHY

The bulk of East Africa comprises a vast and relatively flat plateau that rises from a narrow coastal belt to an average altitude of 1,500m. This plateau is broken dramatically by the Great Rift Valley, which stretches all the way from Mozambique to Arabia. The Rift Valley started to form around 25 million years ago along fault lines associated with tectonic plate activity, the phenomenon that caused the monolithic landmass of Gondwanaland to start separating into its present-day constituent parts some 200 million years ago. Eventually – several million years from now – the Rift Valley is likely to flood completely, and Africa as we know it will split into two or possibly three discrete landmasses.

Two distinct branches of the Rift Valley diverge in a Y shape running northward from the highlands of southern Tanzania. The easterly branch, which runs through the heart of Tanzania and Kenya, was first recognised for what it is by the British geologist John Gregory in 1893, and is still sometimes referred to as the Gregory Rift. The more clearly defined western fork, often referred to as the Albertine Rift, runs along the Congolese borders with Tanzania, Burundi, Rwanda and Uganda to converge with the Nile river system on the Sudanese border. The foot of the Y runs from the southern highlands of Tanzania via Lake Nyasa/Malawi to the Lower Zambezi Valley.

The glacial peaks of Mount Kenya, Africa's second highest mountain, support little life by comparison with the fertile forests of the lower slopes. (AVZ)

The expansion of the Rift Valley has been accompanied by significant volcanic activity, caused by the emergence of subterranean magma between the drifting plates. This violence underlying the Rift's geology is not immediately apparent to the inexpert eye, but on close examination East Africa is studded with plentiful evidence of its unstable past: the sheer stone escarpment that towers over Lake Manyara; the boiling springs that erupt in sulphurous anger alongside Lake Bogoria; the 100-plus crater lakes that lie below the Rwenzoris; and the sheer immensity of the Ngorongoro Crater. Six of Africa's eight highest massifs are situated in East Africa and associated in some way with the rifting process: mounts Kilimanjaro, Kenya, Meru, Elgon and the Virungas are all extinct or dormant volcanoes that formed alongside the Rift within the past 3 million years, while the Rwenzori is a vast geomorphic range that rises from the floor of the Albertine Rift.

Oddly, East Africa is currently one of the most geologically stable areas on earth, and far less prone to earthquakes than the Americas, Asia and Oceania. A rare exception is the southern highlands of Tanzania, where Mount Rungwe last erupted about 200 years ago, and recent tremors measuring up to 4.0 on the Richter scale hint that another eruption is in store. The only live volcano in East Africa is Ol Doinyo Lengai, which has experienced continuous low-key activity since 1883, but the Virungas (on the Uganda/Rwanda border) include two active peaks, both of which

A herd of Uganda kobs in front of the Rift Valley escarpment in Semuliki Wildlife Reserve. (AVZ)

are set a short distance within the Democratic Republic of Congo (DRC). These are Nyamuragira, which is Africa's most boisterous volcano, with more than 40 eruptions documented since 1885, the most recent in 2021; and Nyiragongo, which is the most malevolent, having claimed at least 2,000 lives in the 1977, 2002 and 2021 eruptions.

The Rift Valley floor supports most of Africa's largest lakes, including the serpentine Tanganyika, which ranks as the world's longest and second-deepest freshwater body. The expansive lakes Albert, Kivu and Edward also lie within the Albertine Rift, while the southern Rift is home to the immense Lake Nyasa (known as Lake Malawi in Malawi). The Gregory Rift, by contrast, generally holds far smaller, shallower lakes such as Manyara, Naivasha, Baringo, Natron and Nakuru, a major exception being the vast desert-fringed Lake Turkana in northern Kenya. Biggest of all is Lake Victoria, the world's second-largest freshwater body, which extends over some 68,000km^2 of Tanzania, Uganda and Kenya in an elevated depression situated between the two major forks of the Great Rift Valley, providing food and/or water to at least 30 million people.

Major river systems east of the Rift Valley include the Rufiji/Ruaha, Pangani and Tana, none of which is very large by global standards, reflecting the eastern interior's rather dry climate. Further west, rainfall levels are higher but most rivers drain westward into the Congo River via Lake Tanganyika or northward into the White Nile via lakes Victoria, Edward or Albert. East Africa is the source of the world's longest river, the Nile, which officially flows for 6,650km (4,130 miles) from its most remote headwater, a hillside source of the Kagera River in Burundi. In 2006, the rafting expedition 'Ascend the Nile' located a more remote headwater of the Kagera in Rwanda's Nyungwe Forest, a discovery that will add 100km to the documented length of the Nile if and when it is formally accepted.

A product of recent volcanic activity in northern Rwanda, Lake Burera is a flooded valley dammed by a solidified lava flow from the Virungas. (AVZ)

Elephants under a stormy sky in Tsavo East National Park. (AVZ)

CLIMATE

On the whole, East Africa has a very pleasant climate, more temperate than its tropical location might suggest. However, there are many regional climatic variations, which are influenced by several factors, most significantly altitude. The hottest area is the coast, where typical daytime temperatures of 28–35°C are exaggerated by high humidity and there is little natural relief at night except from sea breezes. Low-lying inland areas such as the Rift Valley floor, in particular around lakes Nyasa/Malawi, Albert and Tanganyika, are also hot, but tend to be less humid and thus more comfortable. At altitudes higher than 1,000m (eg: Arusha, Fort Portal or Nakuru) daytime temperatures are warm to hot, and above 2,000m (eg: Ngorongoro Crater rim) they are moderate to warm. Inland, things tend to cool down significantly at night, and montane areas can be downright chilly after dark. Alpine conditions and sub-zero night-time temperatures are characteristic of the higher slopes of Kilimanjaro and other mountains topping the 4,000m mark.

Because it spans the Equator, East Africa doesn't experience the dramatic seasonal contrast between summer and winter that people are used to in Europe or North America. Overall, October to April is marginally hotter than May to September, with the coast in particular being more pleasant in the cooler months. A more significant seasonal factor is rain, most of which falls between November and May, though the local pattern varies greatly from one place to the next. The rainy season is generally spilt into the short rains or *mvuli* (November and December), and the long rains or *masika* (late February to early May), a pattern that is most marked east of the Gregory Rift. South of Dar es Salaam and west of the Gregory Rift, rain tends to fall fairly consistently from mid-November to mid-April. Uganda and Rwanda are far wetter overall than Kenya and Tanzania, and rain can be expected at any time of year, though the main rainy season again runs from November to May in most parts of the country.

Kenya's Lake Turkana is the world's largest desert lake. (AVZ)

The Batwa people of western Uganda and Rwanda still practise hunter-gathering. (AVZ)

THE HUMAN LANDSCAPE

The timescale of the human occupation of East Africa is unimaginably vast, with something like 300,000 generations separating modern-day safari-goers from their earliest bipedal ancestors, whose fossils litter the area. Modern humans *Homo sapiens* are thought to have evolved within the Rift Valley and have inhabited it for half a million years. For 99% of that term, humankind were exclusive hunter-gatherers, but this most ancient of human lifestyles fell into retreat about 3,000 years ago, following the arrival of the first pastoralists from further north and west. Relict hunter-gatherer societies include the Hadzabe of Tanzania, the Batwa 'Pygmies' of western Uganda and Rwanda, and the hippo-chomping El Molo of Lake Turkana.

Prior to the arrival of Europeans, most of East Africa was inhabited by dedicated pastoralist societies. In drier eastern parts, tribes such as the Maasai, Barabaig and Samburu tended to be strongly nomadic, and to depend almost entirely on livestock for sustenance – in some cases supplemented by hunting. Even today, the most northerly stretch of the Gregory Rift, stretching from the Danakil Depression to Mount Hanang, is inhabited by a near-continuous succession of pastoralist societies that more or less adhere to their ancestral lifestyle.

By contrast, moister parts of pre-colonial East Africa (such as Rwanda and Uganda) supported more sedentary and, in some cases, centralised polities, which practised a mixture of pastoralism and agriculture, with bananas in particular forming an important staple. The first part of East Africa to experience sustained contact with the outside world was the coast, whose Swahili inhabitants became involved in maritime trade more than 1,000 years ago, when ports such as Kilwa, Mombasa and Sofala were regularly visited by Arabian and Asian merchants seeking to load their ships with African produce such as gold and ivory.

A widespread misconception, based on the recent decline of wildlife populations throughout Africa, is that its traditional cultures are inherently non-conservationist. This is demonstrable nonsense. Clearly, it was the arrival of Europeans (who had already triggered numerous extinctions on their home continent, and have since

The Maasai are among several traditional pastoralist groups in the Tanzanian Rift Valley. (AVZ)

Tea estates encroach on Rwanda's Nywungwe Forest National Park. (AVZ)

done so on every other continent they settled) that precipitated Africa's current conservation crisis. The colonial era, though it endured for a mere 75 years, witnessed the large-scale destruction of natural habitats throughout East Africa, along with the indiscriminate slaughter of wildlife herds that had co-existed alongside African people for millennia upon millennia.

By the time of independence, East Africa's larger mammals were more or less restricted to national parks and other conservation areas, most of which were set aside by the colonials without grassroots consultation. As a rule, locals were relocated to the fringes of these newly proclaimed parks, any subsistence hunting they might formerly have practised was criminalised and rebranded as poaching, and they had no legal recourse to prevent wildlife from trampling or eating their crops. A circle of poverty soon grew up around many game reserves, inevitably pitting the local human inhabitants against the wildlife. This scenario was readily exploited by the cash-rich commercial ivory and rhino-horn traders during the poaching wars of the 1980s.

For all this, the prevailing trend in East Africa, at least so far as large mammals are concerned, is rather encouraging, with those national parks and reserves most affected by the poaching of yore having generally experienced a strong growth in most wildlife populations since 1990. Recent decades have also seen conservation authorities move away from the dictatorial attitudes of the colonial and immediate post-colonial eras, and try to involve local people in the process – most notably by encouraging the creation of low-key ecotourism projects that directly benefit local communities.

But while large mammals living in protected areas are in most cases faring tolerably well, the destruction of biodiversity-sustaining microhabitats continues, as does deforestation of many unprotected areas. The main cause of this habitat loss is a rapidly expanding human population, which increased tenfold during the 20th century and shows no sign of abating anytime soon. Still, the good news is that East Africa probably supports more wildlife than any comparably sized area on Earth, and for the most part – influenced partially by an ecotourism boom whose financial impact should never be underestimated – the region's governments seem to be intent on preserving this natural wealth for the benefit of future generations.

HABITATS

Plains zebras in the Ngorongoro Crater. (AVZ)

WHAT IS A HABITAT?

The habitat of any given animal is the place where it naturally lives, as determined by such criteria as climate, vegetation and altitude, along with other ad hoc requirements such as the absence of competing species, the abundance of prey and the availability of suitable breeding sites. Understanding and identifying different habitats will greatly enhance an East African safari. Not only will it help you anticipate what species you're likely to see in any given location, but it can also be an important aid when it comes to distinguishing similar-looking species from one another, and will help build up a holistic picture of the relationship between co-existing animal and plant species.

Some animals have very specific habitat requirements. Hippos, for instance, cannot survive long away from suitable bathing sites, while the range of the aardwolf mirrors that of the termite genera on which it feeds, and most brood-parasitising cuckoos share habitats with their host species. Other creatures, such as the elephant or leopard, are rather less specialised in this respect, though even they have certain physical requirements or preferences. The secretive leopard, for instance, is most common in rocky or bushy areas that offer plenty of cover, while the thirsty elephant must stay within daily walking range of drinking water.

The term habitat might refer to a very specific location, such as a lily-covered pool or boulder-strewn cliff, or to a more generic landscape such as a desert or forest. The grassy Serengeti Plain, a more or less contiguous entity stretching over hundreds of square kilometres, is a good example of a macro-habitat. The isolated rocky outcrops that punctuate this great plain are, by contrast, classic micro-habitats, in that they comprise small self-contained ecological units inhabited by a different community of wildlife to their grassy surrounds. Other habitats have a more cyclical, ephemeral and/or scattered nature, such as the perennial rivers that run through the Serengeti, and the temporary roadside ponds that supplement them during the rains. In all instances, however, anything we describe as a habitat should be a reasonably definable and tangible ecological unit: most lay readers will agree on what constitutes a forest, a pond or a desert, will recognise one when they see one, and would regard the point where one ends and the other begins to be reasonably clear cut.

Elephants consume up to 200 litres of water daily, and are often observed alongside rivers and lakes. (AVZ)

The life-sustaining Ewaso Ng'iro River flows through the arid badlands of Samburu-Buffalo Springs. (AVZ)

HABITAT, ECOSYSTEM OR BIOME?

Individual habitats are the foundations underlying more complex and abstract ecological units such as ecosystems or biomes. An ecosystem, defined by the Convention on Biological Diversity as a 'dynamic complex of plant, animal and micro-organism communities and their non-living environment interacting as a functional unit', typically consists of a mosaic of different habitats. For instance, the grassland, rocky outcrops and aquatic micro-habitats mentioned above all form part of the Serengeti ecosystem, as do a variety of other woodland, riparian forest and montane habitats integrated within the same area. In turn, any given ecosystem, based on its predominant vegetation type, is placed within a biome, a term used to describe all those places on the earth that share a comparable ecosystem.

As an example, Serengeti and Nyerere national parks lie in the same country, protect a similar selection of habitats, and harbour many of the same animal species. But they cannot be regarded as part of the same ecosystem, simply because they lie too far apart to share any meaningful level of wildlife traffic. Furthermore, in ornithological and botanical terms, these two great reserves fall on opposite sides of the boundary separating the acacia-dominated Somali-Maasai biome from the *Brachystegia*-dominated Zambezian biome. Yet if one uses the broadest global definition, the Serengeti and Nyerere can both be placed in the same 'Tropical and subtropical grassland, savannas and shrubland' biome as broadly similar habitats in Asia and South America. Confused? That brings us back to the virtue of simplicity: if you park alongside a permanent pool in the Serengeti or Nyerere, you'll most likely see hippos wallowing in the shallows, sandpipers and storks picking along the shore, and a few antelope ruminating in the shade – different ecosystem, arguably even a different biome, but the habitat remains the same.

SAVANNAH

For most people, the first image conjured up by the mention of East Africa is one of vast grassy plains studded with flat-topped acacia trees and teeming with antelope – a habitat referred to by many African residents as 'bush' and known to ecologists as 'savannah'. The most famous East African savannah habitat is the vast Serengeti-Mara ecosystem, which spans the border of Kenya and Tanzania immediately east of Lake Victoria. This archetypal African savannah comprises 'short grass plains', as well as more densely wooded 'long grass plains' given a golden red hue by the red oat grass *Themeda triandra*. Elsewhere savannah is strictly defined as grassland with scattered trees, but the term is often used to describe any grassy habitat – treeless or wooded, arid or waterlogged – that falls short of being true woodland. The classic African savannah, which consists of a grassy understorey studded with fire-resistant (and mostly deciduous) trees, is regarded by some ecologists as the natural climax vegetation type in areas marked by long and well-defined wet and dry seasons. Others, however, have suggested that the present-day extent of savannah reflects centuries of deliberate burning by pastoralists seeking to stimulate fresh growth to feed their cattle, aided and abetted by the tree-shredding activities of elephants.

SAVANNAH TREES

Highly characteristic of the East Africa savannah are the acacia thorn-trees of the genus *Vachellia* and *Senegalia*. These include the tall umbrella thorn acacia (*V. tortilis*) and flat-top acacia (*V. abyssinica*); the more clumped and shrub-like three-thorned acacia (*S. senegal*) and hook thorn (*S. mellifera*); the swamp-loving, jaundice-

Serengeti National Park protects the world's best-known savannah. (AVZ)

White rhino grazing between yellow fever trees on the shore of Lake Nakuru. (AVZ)

barked yellow fever tree (*V. xanthophloea*); and the scraggly whistling thorn (*V. drepanolobium*), so named for the low whistling sound created by the wind passing through ant galls fashioned around its twinned thorns.

Other more thinly distributed savannah trees include the likes of the sausage tree (*Kigelia africana*), which has a thick evergreen canopy and is often associated with watercourses. The name sausage tree refers to its gigantic pods, which are eaten by elephants and used by the Maasai as gourds. The candelabra tree (*Euphorbia candelabrum*) is a superficially cactus-like succulent with an inverted umbrella shape that grows to more than 10m in height. The dark wood of the African ebony (*Dalbergia melanoxylon*), known locally as *mpingo*, is favoured by the Makonde and other traditional East African carvers.

A male lion casts an imperious eye over the open plains of the Maasai Mara. (AVZ)

Studded with giant euphorbia and boulders, koppies form an important micro-habitat in the otherwise open Serengeti Plains. (AVZ)

SAVANNAH WILDLIFE

The savannahs of East Africa are the favoured habitat of most grass-eating ungulates and support a multitude of grazers, most famously the immense herds of blue wildebeest, common zebra and gazelles that migrate annually across the Serengeti-Mara. Other large grazers associated with lightly wooded savannah include eland, hartebeest, topi, reedbuck and oribi, while more densely wooded areas are favoured by impala, buffalo, giraffe and warthog. The abundance of prey supports plenty of predators, notably lion, spotted hyena and black-backed jackal, with cheetah and serval being more at home in open grassland, and leopard preferring more wooded areas.

Open grassland tends to support a limited avifauna but is notable for the presence of heavyweights such as ostrich, kori and other bustards, secretary bird and ground hornbills, alongside a variety of other ground-dwelling plovers, larks, longclaws, ground barbets and cisticolas. The more densely wooded the savannah, the more varied the birdlife, with open perchers such as the rollers, shrikes, bee-eaters and various raptors occurring alongside the more active sunbirds, lovebirds and other parrots, hornbills and starlings, and less conspicuous bush-shrikes, owls, woodpeckers, cuckoos and batises.

KOPPIES AND CLIFFS

The ancient granite outcrops that so dramatically punctuate the otherwise flat landscape of the southern Serengeti are known locally as koppies (or *kopjes*), from the Afrikaans/Dutch meaning 'little heads'. Studded around several other parts of East Africa, especially in the vicinity of Lake Victoria, these gigantic inselbergs possess something of an island ecology, offering permanent or part-time refuge to a range of plants and animals that couldn't easily survive on the shadeless plains that lap at their base like a grassy ocean.

The most conspicuous koppie dwellers are the bush and rock hyraxes, which often live alongside each other on the rocks, with the former grazing on the grass around the rocks and the latter climbing up to feed mainly on acacia and other trees. Hyraxes form an important part of the food chain in any koppie environment: they are the staple diet of the mighty Verreaux's eagle and other raptors that sometimes nest on the pinnacles, and are also prey for leopards and smaller cats that take daytime refuge in the rocks, as well as pythons that live in the crevices between the giant granite slabs.

Koppies are also favoured by baboons, which for some reason seem to be more skittish and vocal in this rocky environment than they are on the open plains. Antelopes that are commonly associated with koppies are the rock-hopping klipspringers, which seldom leave the mid to upper slopes, and the diminutive dik-diks that lurk around the base. For birders, this is a good place to seek out nightjars (the spectacular pennant-winged nightjar often displays above koppies at dusk), while conspicuous diurnal species include the red-bellied mocking cliff-chat and duller tail-flicking familiar chat (*Oenanthe familiaris*).

Leopards make good use of the cover afforded by rocky outcrops. (AVZ)

The koppies in the Serengeti are among the most spectacular in East Africa, and although most are off-limits to foot traffic, several lodges are constructed around them, allowing for close-up encounters with hyraxes and the brightly coloured agama lizards that bask conspicuously on the rocks. Altogether different, the little-visited koppies that stud the Kondoa region, a day's drive south of Arusha, are home to one of the world's most diverse and prolific natural galleries of prehistoric rock art. Elsewhere, there's nothing stopping tourists from climbing the myriad koppies that rise around Lake Victoria (especially in the vicinity of Mwanza) but do be cautious about where you put your hands and feet: this is prime puff adder territory!

A female greater kudu browses among tangled miombo woodland in southern Tanzania. (AVZ)

WOODLAND

It is difficult to say where savannah ends and woodland begins, especially as the two habitats often occur alongside or through each other. In an African context, however, woodland might best be defined as grassland shaded by a more-or-less continuous tree cover, but lacking the interlocking canopy, sub-canopies and luxuriant undergrowth characteristic of true forest. In northern Tanzania and Kenya, the likes of Tarangire, Manyara, Nakuru and Tsavo contain as much woodland as they do savannah, with various acacia species once again dominant, but often interspersed with baobab trees (*Adansonia digitata*).

East Africa's most extensive woodland habitats lie in southern Tanzania, where Ruaha National Park and environs form the most northerly extent of the so-called miombo belt, which covers most of Zimbabwe, Zambia and Malawi. Miombo woodland typically grows on infertile soil, and is dominated not by thorny, narrow-leaved acacia trees but by broad-leafed, non-barbed trees of the genus *Brachystegia*, often given a rather autumnal appearance by their red and- yellow tinged-leaves.

Woodland habitats tend to protect a greater number and diversity of birds than savannah, with similar families being represented, but mammals are fewer and farther between – or at least seem to be, due to the poorer visibility afforded by the dense foliage. The miombo woodland of southern Tanzania is the most important East African stronghold for roan and sable antelope, greater kudu, African wild dog and side-striped jackal, while typical woodland species found throughout the region include elephant, buffalo, vervet monkey, bushpig and browsing antelope such as Kirk's dik-dik and lesser kudu.

FOREST

The terms forest and woodland, often treated as interchangeable in everyday use, have quite different ecological applications. In theory, the distinction is quite simple: if it has a closed tree canopy it's a forest. In practice, Africa's forests have a very different atmosphere to woodland habitats, and you won't need to study the canopy to know which is which. Put simply, most African forests feel like a tropical jungle, whereas woodland habitats generally don't. The highest forest trees tend to be far taller (up to 45m) than anything you find in broken woodland. They usually overshadow one or more vertically layered sub-canopies, and support a tangle of undergrowth, epiphytes and vines. The impermeable canopy also gives the forest interior a decidedly gloomy and airless quality, which contrasts strongly with the dappled sunlight that illuminates open-canopy woodland.

In terms of land coverage, indigenous forest is a relatively insignificant habitat in East Africa. It accounts for perhaps 1–2% of the total surface area of Kenya and Tanzania. And although it is more profuse in Uganda and Rwanda, up to 90% of these countries' forest has been lost over the past 150 years, either as a result of

A waterfall plunging down in the lush forest of Amani Nature Reserve. (AVZ)

Forest is the favoured habitat of most primates, including chimpanzee in Mahale Mountains National Park, Tanzania. (AVZ)

slash-and-burn clearance, or to make way for sterile plantations of fast-growing conifers and eucalyptuses. Despite this, the forests are East Africa's richest repository of biodiversity, partially as a function of their natural fecundity, and partly due to their inherently fragmented nature, which provides ideal conditions for speciation.

AFRO-MONTANE FOREST

Most East African forests are classified as Afro-montane, a term that encompasses a number of scattered highland habitats running along the eastern half of Africa from Eritrea to Zimbabwe. East Africa lies at the core of this biologically diverse region, with the ancient Afro-montane forests of the Albertine Rift and the Eastern Arc Mountains both ranked among the world's top 20 biodiversity hotspots. Afro-montane forests as a whole protect in excess of 2,500 endemic plant species, together with more than 100 bird and 100 mammal species found nowhere else in the world. The Albertine Rift in particular boasts one of the highest endemism levels of any global biodiversity hotspot, with 550 plant, 40 bird and 35 mammal endemics, as well as many endemic subspecies including the flagship mountain gorilla.

ISLANDS OF ENDEMISM

Often referred to as the African Galápagos, the Eastern Arc Mountains, which include Kenya's Taita Hills and Tanzania's Pare, Usambara, Uluguru and Udzungwa mountains, provide a superb example of the 'island ecology' that moulds the fauna of forest isolates (a subject explored in Jonathan Kingdon's superb *Island Africa*). The 13 ranges of the Eastern Arc formed at least 100 million years ago along a fault line running east of what is now the Rift Valley, and their forest cover has been sustained continuously for 30 million years by moisture blown in from the Indian Ocean. Within the last 10 million years, these forests have become isolated from similar habitats and from each other to form an 'archipelago' of forested islands jutting out from an ocean of low-lying savannah.

This isolation has resulted in an assemblage of endemic taxa with few peers anywhere on the planet, including 16 plant genera (notably *Streptocarpus*, which includes the 'African violets') and 75 vertebrate species. Among them are 'old endemics' such as the giant sengis or elephant shrews, relics of a lineage that was more widespread 20 million years ago, and 'new endemics' whose ancestral stock came from outside the forest. This diversity is today threatened by widespread deforestation and artificial fragmentation. Of the 12 Eastern Arc ranges within Tanzania, only one retains more than half the forest cover it had 150 years ago, while a mere 2% of the original forest remains on the Taita Hills, the only Eastern Arc range to fall outside Tanzania.

The red-throated alethe is one of several elusive Albertine Rift endemics associated with forests in Rwanda and western Uganda. (AVZ)

As with the Eastern Arc, the Albertine Rift forests are truly ancient, having flourished during a succession of prehistoric climatic changes that caused temporary deforestation in lower-lying areas such as the Congo Basin. This antiquity is confirmed by the presence of several species with distinctly non-African affinities – Itombwe owl, Grauer's broadbill (*Pseudocalyptomena graueri*) and Grauer's cuckoo-shrike (*Ceblepyris graueri*), for instance, are each relics of a migrant Asian stock. The Albertine Rift forests are also in rapid decline. Some 2,000 years ago, a belt of uninterrupted forest covered the eastern escarpment of the Albertine Rift from the Rwenzoris to Burundi. The fragmentation of this forest started at the dawn of the Iron Age, when patches were felled to make way for agriculture, but it has accelerated over the past hundred years or so, particularly in densely populated Rwanda.

LOWLAND FOREST

Lowland forest is more widely associated with west Africa than East Africa, though the vast swathe that covers the Congo Basin does nudge into a few parts of western Uganda, notably Semuliki National Park, the only known east African locale for almost 50 bird species endemic to lowland forests. East Africa is also the epicentre of the 'Eastern Africa Coastal Forest' biodiversity hotspot, a narrow and badly degraded band of lowland forest that runs from southern Somalia to Mozambique, intergrading with the Afro-montane forest of the Eastern Arc in the Usambara Mountains north of Tanga. The coastal forest of East Africa is known for its high levels of endemism, which includes 11 mammal and 11 bird species. Among the more important extant forests are those found on Zanzibar and Pemba, respectively home to an endemic monkey and to four endemic bird species. On the Kenyan mainland, the lower Tana River supports two endemic monkey and one endemic bird species, while the Arabuko-Sokoke Forest has three endemic birds.

Forest interiors support hundreds of different butterfly species. (AVZ)

Forest wildlife

Forests are of particular interest to birdwatchers as the main habitat of hundreds of bird species with a limited distribution. Characteristic species include the familiar African grey parrot, the conspicuous and noisy *Bycanistes* hornbills, the colourful turacos, and a miscellany of inconspicuous warblers, secretive thrushes and lookalike greenbuls. This variety is seldom reflected in the number of species identified on a forest walk, as the dense vegetation and tall canopy make it difficult to obtain clear views. The most conspicuous of the larger forest mammals are primates, ranging from the charismatic great apes to the acrobatic colobus and guenon monkeys. Ungulates are less well represented, but the region's larger forests still often harbour elephant, buffalo, bushbuck and at least one duiker species.

The invertebrate diversity of Africa's forests is incalculable. Butterflies are particularly well represented and it's not unusual to see a dozen species fluttering around a small puddle. Less likeable are the biting safari ants that often march along forest paths and will swarm up any non-strategically placed leg, so do watch where you place your feet and ideally tuck your trousers into your socks to prevent them from sneaking in. A knowledgeable local guide is an invaluable asset in the forest, and you'll generally see a lot more along forest margins than in the interior, which means that the best place to start looking for wildlife is normally along any road cut through the forest.

The outsized silvery-cheeked hornbill is among the most conspicuous of forest birds, thanks partly to its raucous braying call. (AVZ)

RIPARIAN FOREST

The highland and lowland forests of East Africa are typically associated with high precipitation, whether that be in the form of rain, mist or a bit of both. Riparian forest, by contrast, is sustained by surface and subterranean water along the banks of rivers, seasonal watercourses and other wetland habitats in otherwise non-forested habitats. Often dominated by the 'yellow fever' acacia and various leafy *Ficus* trees, riverine forests form important corridors for forest wildlife, and often provide safari-goers with an opportunity to see a selection of species they otherwise wouldn't.

A striking example of this phenomenon can be seen in northern Kenya's Samburu-Buffalo Springs complex, whose rocky semi-arid plains are bisected by the ribbon of lush forest that follows the perennial Ewaso Ng'iro River. The Ewaso Ng'iro provides refuge to the likes of bushbuck, waterbuck, elephant and various bird species not normally observed in this region. Other good examples include those that follow the Grumeti and Mara rivers through the Serengeti-Mara, where black-and-white colobus monkeys and Ross's turaco are commonly seen.

Low mammal densities and otherworldly vegetation characterise the Afro-Alpine moorland belt of the region's larger mountains. (AVZ)

HIGHLANDS

At higher altitudes, the forested lower slopes of East Africa's taller mountains typically give way to an ethereal, pastel-shaded cover of open montane grassland or Afro-alpine moorland. Neither habitat is home to a great variety of animal species, though montane grasslands often host a wealth of flowering perennials such as orchids, proteas, geraniums, lilies and aster daisies. The most substantial East African montane grassland is the 273km² Kitulo Plateau in southern Tanzania, whose rain-drenched volcanic soils support 350 species of flowering plant, including 60-odd Tanzanian endemics. Conspicuous in most montane grassland habitats are the aloes, spike-leafed succulents whose long-stemmed nodular red and yellow flowers typically bloom during the dry season alongside the marsh-loving red-hot pokers, attracting prodigious numbers of colourful sunbirds.

On major mountains such as Kilimanjaro, Kenya and the Rwenzori, a belt of Afro-alpine moorland typically occurs between the 3,000m and 4,000m contours. Generally accessible only by foot (a notable exception being Ethiopia's Bale Mountains), this habitat is characterised by grey-pink heathers studded with otherworldly giant forms of *Lobelia* and *Senecio*, sometimes growing up to 5m high. The moorland zone supports a low density of mammals, but rock hyrax and klipspringer are quite common on rocky outcrops, and eland and elephant wander up from time to time. A limited range of birds includes bearded vulture (*Gypaetus barbatus*), Verreaux's eagle, alpine chat (*Pinarochroa sordida*), scarlet-tufted malachite sunbird (*Nectarinia johnstoni*), alpine swift and various members of the crow family.

Above 4,000m, rainfall tends to be very low and ground temperatures show intense fluctuation, dropping below 0°C at night but soaring above 30°C by day. The few plant species to survive in these demanding conditions are mostly lichens or grasses, and wildlife is thin on the ground – though a frozen leopard was discovered above the 5,000m contour of Kilimanjaro in 1926, and a pack of African wild dogs was recorded in the early 1960s.

SEMI-DESERT

Nowhere in East Africa meets the most stringent ecological definition of a desert, but much of the region – the northeastern two-thirds of Kenya, for instance, and large tracts of central Tanzania – is classified as semi-arid or arid, receiving an annual rainfall of below 500mm. Too dry to support cultivation or commercial ranchland, these zones typically support a scrubby tangle of acacia thicket that transforms into a blanket of flowering greenery after good rains, while the more arid likes of the Chalbi and Dida Galgalu 'deserts' are practically bereft of vegetation. Such aridity is not to everybody's tastes, but many – this writer included – would regard northern Kenya in particular to be among the most compelling parts of East Africa, above all for its endless horizons and liberating sense of untrammelled wild space.

Wildlife tends to be thinly distributed in dry-country areas and comprises mainly desert-adapted creatures that can last for relatively long periods without drinking. Typical mammals include oryx, gerenuk, Grevy's zebra and ground squirrel, though

The handsome reticulated giraffe is confined to the arid 'badlands' of northern Kenya and Somalia. (AVZ)

a greater variety of wildlife is usually found close to springs or rivers, such as those in Kenya's popular Samburu-Buffalo Springs reserves. The acacia scrub that covers much of the region is surprisingly productive for birds and supports numerous regional dry-country endemics, including the spectacular vulturine guineafowl, Somali ostrich, golden pipit and golden-breasted starling. The empty plains of Dida Galgalu and Chalbi support a very limited selection of wildlife but are notable for the presence of a few localised endemic larks.

WETLANDS AND WATERWAYS

Ranging from the sandy beaches of Lake Victoria and steep rocky shores of Lake Tanganyika, to the shallow soda lakes of the Gregory Rift and impenetrable papyrus swamps of southern Uganda, East Africa is richly endowed when it comes to wetlands – a generic term embracing any habitat that combines terrestrial and aquatic features. The significance of wetlands is almost impossible to overstate, not only as a self-sustaining ecosystem supporting its own distinctive wildlife, but even more so as a source of vital drinking water to the inhabitants of most terrestrial ecosystems. Yet this life-sustaining habitat is also frequently threatened by development, whether that be swamp drainage, industrial pollution, or disruption by reservoirs or hydro-electric schemes. Fortunately, East Africa's wetlands are generally in better shape than most. Indeed, Lake Tanganyika has the lowest pollution levels of any major lake in the world. Unfortunately Lake Victoria has fared rather less well, primarily due to the colonial-era introduction of the predatory Nile perch, which has guzzled many of the lake's endemic cichlid species to extinction, and the high levels of pollution from lakeshore industry and settlement.

Mammals that occur more or less exclusively in aquatic habitats include the hippo, sitatunga, marsh mongoose and three species of otter, but the nominally terrestrial likes of elephant and buffalo also regularly take to water, and most other species visit a source of drinking water daily. More than 100 East African bird species are strongly associated with water, ranging from the swallows and martins that feed above it, to the waders that peck in its shallows, and the aerial anglers such as the pied kingfisher

Saline lakes such as Lake Nakuru often support flamingos in their thousands. (AVZ)

Hippos are common in larger rivers such as the Rufiji in southern Tanzania. (AVZ)

and African fish eagle that plunge in to catch their prey. Some water-associated birds, such as the Egyptian goose or western cattle egret, might be seen in practically any aquatic habitat, while others have more specific requirements. The African finfoot and white-backed night heron, for instance, frequent still or sluggish waters with overhanging vegetation, whereas the African skimmer and white-crowned lapwing prefer exposed sandbanks. Quiet lily-covered pools are the favoured haunt of the African jacana, African pygmy goose and long-toed lapwing, while grebes, pelicans and most ducks tend to swim openly on exposed water bodies. East Africa forms an important habitat for several localised papyrus endemics, notably the brightly coloured papyrus gonolek and unique shoebill.

The handsome black-winged stilt is associated with shallow lake margins. (AVZ)

ARTIFICIAL HABITATS

As is the case elsewhere in the world, a significant area of East Africa now supports manmade habitats, ranging from reservoirs and plantation forests to urban settlements and farmland. This isn't necessarily bad news for all wildlife. Artificial reservoirs create fresh habitats for water-associated birds and amphibians, urban settlements often attract unnaturally high densities of hooded vulture and marabou stork, and disturbed farmland can provide ideal conditions for certain seedeaters and finches. But let's not get ahead of ourselves! These examples are very much the exception, since all artificial habitats are created at the expense of natural vegetation, resulting in the partial or total loss of an indigenous habitat, and the local extinction of associated plant and animal species.

In East Africa, the level of habitat loss is most striking in Rwanda, a densely populated and highly fertile country that barely supports a hectare of natural vegetation outside of its three main conservation areas. But the same is true of most medium to high rainfall parts of Uganda, Kenya and Tanzania: what used to be moist savannah is now a cultivated monoculture; what used to be lush indigenous forest is now a sterile plantation of water-table-draining exotics; what used to be an aloe-studded grassy slope is now a subsistence *shamba* planted with papayas, bananas and coffee... and so on. True, the more arid parts of Tanzania and Kenya are less overtly affected by human activity. Even so, many arid areas have been seriously overgrazed as a result of livestock overpopulation, leading to widespread erosion and causing wildlife to starve to death.

A related threat to the diversity and ecological integrity of any given ecosystem is the actual and potential spread of exotics (ie: species that don't occur there naturally). These are sometimes animals, such as the Indian house crow (*Corvus splendens*) that breeds prolifically in several coastal urban centres. More often, however, they are plants that have been introduced deliberately for cropping, plantation or hedging purposes. In many cases, such plants pose no further threat to the environment, because they are poorly adapted to local conditions and cannot survive long or

Marabou storks have adapted readily to the spread of humanity, finding easy pickings around rubbish dumps, slaughterhouses and beaches used by fishermen. (AVZ)

propagate without human intervention. But a small proportion of exotics find local conditions to their liking, and these adaptable aliens – referred to as invasive species – often spread like the proverbial wildfire because they are not kept in check by the natural enemies that control them in their country of origin. The spread of invasive exotics has become a major global issue over the past hundred years, resulting in diminished biodiversity almost everywhere in the world. East Africa is no exception, the famous example being a South American water hyacinth that was introduced to Kigali (Rwanda) as an ornamental pond plant, spread via the Kagera River to reach Lake Victoria in 1989, and subsequently colonised vast tracts of the lake surface, forming an impenetrable mat that clogged up several harbours and leading to a marked depletion in oxygen levels.

It's a sorry state of affairs, and the situation is likely to deteriorate further so long as the human population keeps growing at its present rate. It should, however, be noted, and somewhat emphatically, that East Africa is far less seriously ecologically compromised than most other parts of the world, which is why so much wildlife remains, and why there is still considerable cause for hope.

The hills around Lake Kivu in Rwanda, like many areas, have lost their indigenous forest to tea and other plantations. (AVZ)

MAMMALS

An alert young male lion in Ruaha National Park. (AVZ)

For most first-time visitors to Africa, it's all about the large mammals. Of course we all enjoy the colourful birds, iridescent butterflies and beady-eyed lizards that flap, flutter and flit their way through the tangled bush. But East Africa's biggest drawcard, undoubtedly, is its status as the world's greatest remaining refuge for what biologists refer to, only slightly facetiously, as the sexy mega-fauna. From the gorillas that hide in the misty Virungas to the chimps that patrol the forested shores of Lake Tanganyika; from the glamour of the regal lion and stealthy leopard to the extraordinary extremes of the stalk-necked giraffe and python-trunked elephant; from the graceful herds of gazelle and impala to the stiff-tailed warthog and much maligned hyenas; from snorting hippos and crosspatch rhinos to the faintly

The mountain gorilla, for all its formidable reputation, is a peaceable vegetarian. (AVZ)

preposterous wildebeest and positively ludicrous aardvark – well, goodness, it really does feel as if all mammalian life is to be found here!

Mammals (order Mammalia) are distinguished from other vertebrates by the presence of epidermal hair or fur (as opposed to scales or feathers), but they also share several other common characteristics, including the female mammary (milk-producing) glands for which the order is named, a unique trio of minute bones (ossicles) within the ear, and a neo-cortex in the brain (associated with higher functions such as communication and sensory perception). The first mammals appear on the fossil record around 130 million years ago in the form of undistinguished shrew-like creatures doomed to spend the first half of their earthly tenure cowering in the shadows of the larger and more rapacious dinosaurs. It was the sudden mass extinction of dinosaurs some 65 million years ago that triggered the evolutionary explosion that led to the terrestrial dominance of mammals today.

Allowing for taxonomic vagaries, something like 450 mammal species (or 10% of the global total) are known from East Africa, including well over 100 that are loosely classified as 'large mammals'. All are placed in the subclass

Placentalia, which means that the foetus is fed in the womb with a placenta (as opposed to the egg-laying monotremes and the pouched marsupials, which are confined to Australasia and South America). Not including marine mammals, East Africa's mammalian fauna is conventionally split across 15 orders, of which five contain the large mammals that generally dominate the safari agenda: primates, carnivores, perissodactyls (zebras and rhinos), artiodactyls (buffalo, antelope, hippos, pigs, giraffes) and proboscids (elephants).

It should be noted that the long-standing conventions of mammal classification, based on principles established by Simpson in 1945 and universally accepted for the subsequent five decades, have been brought into question by recent molecular studies. As a result, two very different new reclassifications of the Mammalia have been published since 1995, both of which are somewhat controversial and are largely ignored in the information that follows.

PRIMATES

Seasoned safari-goers tend to underestimate Africa's primate diversity, which probably stands at around 100 species, depending on various unresolved taxonomic ambiguities. This is because primates are essentially creatures of the treetops, and are better adapted to the western rainforest habitats than to the eastern savannahs that form the focus of most safaris. Even so, East Africa supports dozens of ape, monkey and bushbaby species, and it is something of a mystery why – gorillas and chimps aside – this aspect of the region's wildlife is so often underplayed by the travel media and tourist industry. Most East African monkeys are beautifully marked, and all possess a plethora of captivatingly human qualities. As for their dashing acrobatics, well who could watch a troop of black-and-white colobus monkeys leaping with abandon from tree to tree and fail to experience a vicarious thrill at their graceful athleticism?

East Africa's most popular primates are the mountain gorillas of the Rwanda/ Uganda border area, and the chimpanzees that have been habituated for tourist visits at half a dozen national parks in Uganda, Rwanda and western Tanzania. Far more widespread and diverse, however, are the Old World monkeys of the family Cercopithecidae, which are split between two subfamilies, Cercopithecinae (cheek-pouched monkeys) and Colobinae (colobus monkeys). These diverged from other monkeys perhaps 10 million years ago. Also well represented in the region are the nocturnal pottos and bushbabies, which are more closely related to the lemurs of Madagascar than to other African primates.

APES

The great apes share more than 95% of their genes with people, and most taxonomists now place us all in the same family (Hominidae). Indeed, humans and chimps are so closely related that a less partial observer might well place us in the same genus. Four of the world's non-human ape species are endemic to equatorial Africa, with the eastern gorilla and common chimpanzee both present in East Africa, though neither is likely to be encountered casually on a standard safari package.

A bulky silverback on the forest edge of Volcanoes National Park. (AVZ)

Nevertheless, East Africa is unequivocally the best place to see Africa's wild apes. Uganda's Bwindi Impenetrable National Park and Rwanda's Volcanoes National Park reliably offer thrilling close-up encounters with the charismatic mountain gorilla, while the remote Gombe and Mahale Mountains national parks in western Tanzania offer the world's best chimpanzee tracking, followed closely by the likes of Kibale National Park, Queen Elizabeth National Park and several forest reserves in Uganda, and Nyungwe National Park in Rwanda. Many visitors to Uganda also visit Ngamba Island Chimpanzee Sanctuary, which is an admirable set-up, but offers an experience more akin to a zoo than to tracking wild chimps.

Mountain gorilla

Visiting the endangered mountain gorillas of Rwanda and Uganda is widely – and justifiably – regarded to be one of Africa's ultimate travel highlights, and certainly it is difficult to think of a more emotionally uplifting or satisfying wildlife encounter than gazing into the intelligent liquid brown eyes of a mighty silverback as it chomps peacefully on a stem of bamboo. Endemic to the forests of equatorial Africa, gorillas are the world's largest living primates, and among the most peaceable and sedentary of creatures. Sadly, these gentle giants are also known to be endangered throughout their range, with a global population of fewer than 100,000 individuals concentrated mostly in the Congo Basin.

Prior to 2021, all gorillas were assigned to the same species, but DNA tests have since revealed that the western and eastern populations diverged some 2 million years ago. They are now treated as two discrete species, *Gorilla gorilla* (western) and *Gorilla beringei* (eastern), with the latter further split into the Congolese lowland subspecies (*G. b. graueri*) and the mountain gorilla (*G. b. beringei*) of Uganda and Rwanda. The eastern gorilla is the rarer species: the population of lowland gorillas, estimated at 17,000 in the early 1990s, has dropped to 5,000 as a result of the ongoing Congolese civil war, while the mountain gorilla is represented by an estimated 1,060 individuals in the Virunga Volcanoes and Bwindi Impenetrable National Park, according to the most recent census in 2018.

The mountain gorilla was formally described in 1903, a year after two individuals were shot on Mount Sabinyo by Oscar van Beringe. It is distinguished from other gorillas by several adaptations to its high-altitude home, most visibly a longer and more luxuriant coat. The first behavioural study was undertaken in the 1950s by George Schaller, and formed the starting point for the research initiated a decade later by Dian Fossey, author of the landmark *Gorillas in the Mist*. The unsolved murder of Fossey at her research centre in December 1985 is thought to have been the handiwork of one of the many poachers with whom she crossed swords. This massive ape is bulkier than other subspecies of gorilla, with the heaviest individual ever measured being the 220kg dominant silverback of Rwanda's Sabinyo Group. Like other gorillas, it is a sociable creature, moving in defined troops of anything from five to 50 animals. A troop typically consists of a dominant silverback male, possibly a subordinate silverback, a harem of three or four mature females, and several young animals. Unusually for mammals, the male is the focal point of gorilla society; when a silverback dies, his troop normally disintegrates. A silverback will start to acquire his harem at about 15 years of age, most normally by attracting a young sexually mature female from another troop. He may continue to lead a troop well into his forties.

A female reaches sexual maturity at the age of eight, and will often move between several different troops until she has given birth, after which she normally stays loyal to the paternal silverback until he dies and will even help defend him

Adult mountain gorillas are strictly terrestrial but youngsters often ascend clumsily into the trees and bamboo. (AVZ)

against other males. (When a male takes over a troop, he generally kills all nursing infants to bring the mothers into oestrus more quickly, a strong motive for a female to help preserve the status quo.) A female gorilla has a humanlike gestation period, and will typically raise up to six offspring to sexual maturity in her lifetime, assuming that she reaches old age. Her status is based on how long she has been with a silverback: the alpha female is normally the longest-serving member of the harem.

The mountain gorilla is primarily vegetarian. Bamboo shoots form the bulk of its diet, but it is known to eat 58 other plant species in the Virungas, as well as insects for a protein supplement. A troop spends most of its waking hours on the ground, but moves into the trees at night, when each individual builds itself a temporary nest. Surprisingly sedentary creatures, gorillas typically move less than 1km in a day, which makes it easy to locate them on a day-to-day basis. A troop will generally only move further after a stressful incident, for instance an aggressive encounter with another troop. Gorillas have few natural enemies and they often live for up to 50 years in the wild, but their long-term survival is threatened by poaching, deforestation and exposure to human-borne diseases.

Just like humans, chimpanzees have unique facial features that help them recognise each other. (AVZ)

Common chimpanzee

You'll hear them before you see them: from somewhere deep in the forest, an excited hooting, just one voice at first, then several, rising in volume, tempo and pitch to a frenzied unified crescendo, before stopping abruptly or fading away. Jane Goodall called it the 'pant-hoot' call, a kind of bonding ritual that allows any chimpanzees within earshot of each other to identify exactly who is around at any given moment, through the individual's unique vocal stylisation. To the human listener, this eruptive crescendo is one of the most spine-chilling and exciting sounds of the rainforest, and a strong indicator that visual contact with man's closest genetic relative is imminent.

It is, in large part, our close evolutionary kinship with the common chimpanzee (*Pan troglodytes*) that makes this sociable black-coated ape so enduringly fascinating. Humans, chimpanzees and bonobos (also known as pygmy chimpanzees) share some 97% of their genetic code, and are more closely related to each other than to any other living creature, even gorillas. Superficial differences notwithstanding, the

The chimpanzee and bonobo are humans' closest living relatives. (AVZ)

similarities between humans and chimps are consistently striking, not only in the skeletal structure and skull, but also in the nervous system, the immune system and in many behavioural aspects: bonobos, for instance, are the only mammals other than humans to copulate face-to-face.

Unlike most other primates, chimpanzees don't live in troops, but extended communities of up to 100 individuals, which roam the forest in small, socially mobile subgroups that often revolve around a few close family members. Male chimps normally spend their entire life within their birth community, but females regularly migrate to a neighbouring community after reaching adolescence. A high-ranking male will occasionally attempt to monopolise a female in oestrus, but the more normal state of sexual affairs in chimp society is non-hierarchical promiscuity. Young oestral females tend to mate with any male they fancy, but older females often form close bonds with a few specific males, sometimes allowing themselves to be monopolised by a favoured suitor for a period, but never pairing off exclusively in the long term.

Within each community, one alpha male is normally recognised, though coalitions between two males, often a dominant and a submissive sibling, are not unusual. The role of the alpha male is evidently quite benevolent – chairman of the board rather than crusty tyrant. This is probably influenced by the alpha male's relatively limited reproductive advantages over his potential rivals, most of whom he will know for his entire life. Within a community, lower-ranked males are generally supportive of the alpha male, except when a rival consciously contests the alpha position, which is far from being an everyday occurrence – indeed, one male in Mahale Mountains maintained his alpha status from 1979 to 1995!

Prior to the 1960s, it was always assumed that chimps were strict vegetarians. This notion was rocked by Jane Goodall, who first witnessed chimps hunting down a red colobus monkey in the early years of her residency at Gombe National Park. Hunting is now known to be common behaviour, particularly during the dry season when other food sources are depleted. Over subsequent years, an average of 20 kills has been recorded in Gombe annually, most frequently red colobus, but young bushbuck, young bushpig and even infant chimps have also been taken. The normal modus operandi is for four or five adult chimps to encircle a colobus troop, then for another chimp to act as a decoy, creating deliberate confusion in the hope that it will drive the monkeys into the trap, or cause a mother to drop her baby.

Chimp communities are reasonably stable entities, but intra-community warfare has been recorded in Mahale and Gombe. In Mahale, one of the two communities habituated by researchers in 1967 had exterminated the other by 1982. A similar thing happened in Gombe National Park when the community habituated by Goodall divided into the discrete Kasekela and Kahama communities. The two communities co-existed peacefully for some years, but then suddenly the Kasekela males started to persecute their former allies, isolating them one by one, and tearing into them until they were dead or fatally wounded. By 1977, the Kahama community had vanished entirely.

Chimpanzees are essentially inhabitants of the western rainforest, but their range extends into Rwanda, Uganda and western Tanzania, which have a combined population of 6,000–8,000 individuals. These are concentrated in Tanzania's Mahale and Gombe national parks, Rwanda's Nyungwe Forest, and about 20 Ugandan national parks and other reserves, most notably Budongo, Kibale, Kalinzu, Maramagambo and Bwindi. Although East Africa's chimps represent less than 5% of the global population, much of what is known about wild chimpanzee society and behaviour stems from the region, in particular the ongoing research projects initiated in Gombe and Mahale Mountain national parks back in the 1960s.

An interesting pattern that emerged from the parallel research projects in these two reserves is the strong behavioural differences between their chimp populations. For instance, while Gombe's chimps regard the palmnut to be something of a delicacy, their counterparts at Mahale have yet to be observed eating the same fruit, though it grows profusely in the park. Likewise, the famed 'termite-fishing' behaviour recorded at Gombe National Park in the 1960s has a parallel in Mahale, where the chimps are often seen 'fishing' for carpenter ants in the trees. But the Mahale chimps have never been known to fish for termites, while the Gombe chimps evidently do not fish for carpenter ants. More, arguably, than any other aspect of chimp society and behaviour, it is these cultural differences – the influence of nurture over nature if you like – that bring home their close genetic kinship to humans.

Chimpanzees set out on a foraging trip. (AVZ)

COLOBUS MONKEYS

The thumbless, leaf-eating, forest-dwelling colobuses (family Colobidae) are easily distinguished from other African monkeys by their rather elegant combination of small head, very long tail and spidery limbs. All colobus monkeys are strongly arboreal, rarely venturing to the ground, and they subsist almost entirely on plant matter, which is processed by a somewhat ruminant-like digestive system. Two distinct genera are recognised in East Africa, with *Colobus* being represented by three species of pied colobus and *Piliocolobus* by four localised species of red colobus.

Pied colobuses

One of the most striking primates in East Africa is the eastern black-and-white colobus (*Colobus guereza*), whose luxuriant black coat contrasts strongly with its bold white facial markings, sides and shoulders, and the long white tail that streams behind it spectacularly when leaping up to 30m between trees. Almost exclusively arboreal, this handsome monkey can weigh up to 12kg and grows to be 65cm long, excluding the tail. It is typically seen in parties of up to ten adults, sometimes with a couple of albino-like babies in two. This species is common in many mid-

The striking eastern black-and-white colobus is the most common forest monkey in much of East Africa. (AVZ)

altitude and highland forests, and particularly easy to see in Aberdares National Park (Kenya) and Entebbe Botanical Garden (Uganda). As of 2018, the subspecies known as Kilimanjaro guereza (*C. g. caudatus*), a Tanzanian near-endemic confined largely to forests on Kilimanjaro and Mount Meru, is sometimes regarded as a full species.

The eastern black-and-white colobus is replaced by the similar but less extravagantly tailed Angola colobus (*C. angolensis*) in forested habitats along the Indian Ocean coastline and the Eastern Arc Mountains, as well as in parts of the Albertine Rift. Arguably the best colobus watching in East Africa is to be had in Rwanda's Nyungwe National Park, where a single habituated troop of 500-plus Ruwenzori colobus (*C. a. ruwenzori* – a subspecies endemic to the Albertine Rift) is thought to be the largest arboreal primate troop anywhere in Africa.

Female Zanzibar red colobus nursing young. (AVZ)

Red colobuses

Generally more nondescript than their pied relations, East Africa's four red colobus species all have some black on the upper back, red on the lower back, a pale tufted crown and a long-limbed appearance unlike that of any guenon or mangabey. All four species are IUCN listed as Endangered or Critically Endangered, the most widespread among them being the Ugandan red colobus (*Piliocolobus tephrosceles*), which occurs in several forests in western Tanzania and Uganda, most notably Kibale National Park, where it is quite commonly seen along the main access road.

The most striking species is the Zanzibar, or Kirk's, red colobus (*P. kirkii*), whose unkempt white crest calls to mind a colobine Einstein! Endemic to Zanzibar Island, this monkey's main stronghold is Jozani-Chwaka Bay National Park, where a few habituated troops are easily observed at close quarters. It is reasonably well protected within its limited range, and an estimated population of more than 5,000 individuals makes it considerably more secure than that of the Udzungwa red colobus (*P. gordonorum* – endemic to the Udzungwa Mountains) and Tana River red colobus (*P. rufomitratus* – endemic to the Tana River Delta), neither of which numbers more than 1,000 in the wild.

CHEEK-POUCH MONKEYS

The cheek-pouch monkeys of the Cercopithecidae are a more varied bunch than their colobine cousins, and occupy a greater range of ecological niches, from dry savannah and montane moorland to evergreen forest. They are named for the characteristic pouch in their inner cheek, which in certain species can hold as much food as a full stomach. The guenons of the genus *Cercopithecus* are represented by around ten forest-associated species in East Africa, but the savannah-dwelling baboons and vervet monkey are more likely to be encountered by casual safari-goers.

The vervet monkey is equally at home on the ground or in the trees. (AVZ)

Vervet monkeys make devoted mothers. (AVZ)

Vervet and patas monkeys

The vervet monkey (*Chlorocebus pygerythrus*) is possibly the world's most numerous primate (apart from humans), and is distributed throughout eastern Africa, from the Cape to southern Ethiopia. It is taxonomically controversial, but five subspecies are now recognised, while most authorities regard the similar green and grivet monkeys of west and northeast Africa as distinct species. Their typical coloration is grizzled light olive or grey off-set by a black face, white ruff, pale belly and superbly unapologetic bright blue scrotum.

Weighing in at around 5kg, the vervet is a common resident of savannah, open-canopy woodland, forest fringe and suburbia, where troops of up to 75 individuals spend a high proportion of their time on the ground, squabbling, playing and foraging for fruit, seeds, gum and insects. The vervet is generally unmistakable, though in certain dry-country areas (northern Serengeti, Murchison Falls National Park and along the Kenya/Uganda border) it occurs alongside the more spindly patas monkey (*Erythrocebus patas*). This superficially similar species has ganglier limbs and a russet-tinged coat. It lives almost entirely on the ground, and is known as the fastest of all primates, able to run at speeds of more than 50km/h.

A patas monkey keeps a lookout. (AVZ)

A yellow baboon feeding on a fruit. (AVZ)

The olive baboon is darker in colour than the yellow. (AVZ)

Baboons

Baboons are among the most terrestrial of primates, and often occur alongside vervet monkeys in savannah habitats. They can be distinguished from all other East African monkeys by their greater bulk (up to 45kg), inverted U-shaped tail and distinctive doglike head. Baboons are fascinating to watch from a behavioural perspective, living in large, quarrelsome and shag-happy troops that boast a complex but rigid social structure, characterised by matriarchal lineages and plenty of intra-troop movement between males seeking social dominance.

A versatile omnivore, the baboon is at home in almost any habitat. It is frequently seen in most East African game reserves, as well as in mountainous areas, where hikers will often first be made aware of a troop's presence by their far-carrying barking calls. Baboons generally steer clear of people, but can become aggressive where they see us as a potential source of a free meal – something that occasionally happens around campsites, such as at Lake Nakuru National Park (Kenya). You should treat any such baboons with considerable caution.

48

Two species of baboon are present in East Africa. The bulkier and darker green-brown olive baboon (*Papio anubis*) is mainly distributed to the west of the Rift Valley while the more lightly built yellow-brown yellow baboon (*P. cynocephalus*) occurs in the far south of the region and to the east of the Rift Valley.

Forest guenons

The enchanting and sociable guenons of the genus *Cercopithecus* are medium-small monkeys, most of which have attractive but rather cryptically hued markings. They are the most taxonomically confusing of all African primate groupings: anything from 10–35 species are recognised, more than 100 subspecies have been described, and the picture is further complicated by a strong tendency towards hybridisation between apparently distinct species. This high level of genetic instability probably reflects a vast expansion in the range of prototypal guenons over the past 5 million years, resulting in the largest known explosion of radial divergence in primate history. The mind-boggling complexities of guenon taxonomy necessitate a fair degree of simplification in the species accounts that follow:

Blue monkey

The most widespread forest guenon is the blue monkey *Cercopithecus mitis*, whose range – extending from the east coast of South Africa to the Ethiopian highlands and Congo Basin – reflects its willingness to follow riverine woodland and similar corridors through savannah habitats. Primatologists recognise up to two dozen subspecies, a level of regional variability reflected in the local use of half a dozen other vernacular names, notably Sykes', samango, gentle, silver, white-throated and diademed monkey. Some regard Sykes' monkey (*C. m. albogularis*) as a separate species.

Blue monkeys generally live in small troops, and might be encountered in any suitably forested habitat in East Africa. Most subspecies have a uniform dark blue-grey coat broken by a white throat (sometimes extending down the chest or around the collar), making them readily distinguishable from any race of pied

Sykes' monkey is a widespread race of blue monkey in Kenya and northern Tanzania. (AVZ)

or red colobus. Reliable spots for blue monkey sightings include the forested slopes of Ngurdoto Crater in Arusha National Park, the groundwater forest near the entrance of Lake Manyara National Park, and Mountain Lodge on the southern slopes of Mount Kenya.

Sometimes still treated as a subspecies of blue monkey (*C. (m.) kandti*), the somewhat contrarily named golden monkey is a bamboo-associated Albertine Rift endemic, whose orange-gold body, cheeks and tail contrast strikingly with its black limbs, crown and tail tip. Formerly more widespread, this pretty monkey

The golden monkey is a race of blue monkey confined to bamboo forest. (AVZ)

is now near-endemic to the Virunga Volcanoes, where it is the most common primate. Several golden monkey troops have been habituated to daily tourist visits in Rwanda's Volcanoes National Park or Uganda's Mgahinga Gorilla National Park. Tracking these charming monkeys makes a great add-on to gorilla tracking in the same parks.

The red-tailed monkey has a distinctive nose. (AVZ)

Red-tailed monkey

The red-tailed monkey (*C. ascanius*) is a widespread and hyperactive forest inhabitant. Its brown coloration is off-set by a coppery-red tail and a distinctive heart-shaped white nose-blob that vaguely recalls an albino version of the red nose worn for Comic Relief! A mixed frugivore and insectivore, this small and highly arboreal monkey is normally seen singly or in pairs, but it also associates with other monkeys, and aggregations of 100-plus have been recorded at seasonal food sources. It occurs in most large forests in the western part of the region, including Kakamega (Kenya), Nyungwe (Rwanda), Kibale, Bwindi and Mpanga (Uganda), and Mahale and Gombe (Tanzania). It sometimes hybridises with the blue monkey in Kibale National Park.

L'Hoest's monkeys are common in Rwanda's Nyungwe Forest. (AVZ)

L'Hoest's monkey

Among the largest of the forest guenons, clocking in at up to 10kg, L'Hoest's monkey (*Allochrocebus lhoesti*) is unusually terrestrial, feeding in the trees but generally travelling between feeding grounds along forest-floor paths. It is more or less confined to the Albertine Rift, where it is rendered unmistakable by the combination of its large size, habitually upright tail and cryptic grey-red coat off-set by a bold white 'beard'. This species is the most visible monkey in Rwanda's Nyungwe Forest, with troops of 5–15 animals often seen along the road and around the campsite. In Uganda, it is present in Kibale and Maramagambo forests, where sightings are relatively scarce, but it is quite common in Bwindi Impenetrable National Park along the road connecting Buhoma to the Waterfall Trail.

Other guenons

De Brazza's monkey (*Cercopithecus neglectus*), an impressive thickset monkey with a distinctive white moustache and beard, is a western species whose East African range is restricted to Semuliki National Park and Kenya's Saiwa Swamp, where it is quite easily observed on foot. Also very localised in East Africa, Dent's mona (*C. denti*) is confined to Rwanda's Nyungwe Forest, where it often moves with other monkey species but can be distinguished by its contrasting black back and white belly, blue-white forehead, and yellowish ear tufts. The secretive owl-faced or Hamlyn's monkey (*C. hamlyni*), a thickset, pug-faced inhabitant of bamboo forests in the eastern DRC and Albertine Rift, occurs locally in southwest Uganda and Rwanda's Nyungwe Forest, but it is highly unlikely to be seen by casual tourists.

Uganda mangabey (AVZ)

Mangabeys and allies

Intermediate in size between baboons and guenons, mangabeys are dull-coloured arboreal monkeys associated primarily with low-altitude rainforest in central Africa. Of the western species, only the grey-cheeked mangabey (*Lophocebus albigena*) has a range nudging into East Africa, where it is quite common in Nyungwe National Park. The Uganda mangabey (*L. ugandae*), often regarded as a subspecies of the above, is a Uganda endemic most easily seen in Kibale National Park. Both are distinguished from other forest monkeys by a spindlier appearance and loud gobbling call.

Oddly, however, Kenya and Tanzania are home to two very localised endemic mangabey species, whose ranges are isolated from each other by about 1,000km. Both are relatively terrestrial inhabitants of riparian forest, and probably form relict populations of a 'forest corridor' genus displaced elsewhere in East Africa by guenons and baboons. Of the two, the Tana mangabey (*Cercocebus galeritus*) warrants the more immediate conservation concern, its range restricted to a 60km belt of woodland following the lower Tana River through eastern Kenya. Despite nominal protection, this species is listed as Critically Endangered by the IUCN with a population of fewer than 1,000 individuals.

By contrast, the existence of the Sanje crested mangabey (*C. sanjei*) was unknown to Western science in 1972. It was discovered by Katherine Homewood, who heard a distinctly mangabey-like whoop from within the forest whilst she was studying red colobus at Sanje Waterfall (Tanzania) in 1979. At first Homewood thought she was hallucinating or the victim of a practical joke, but her guide recognised the call as that of a monkey known locally as n'golaga and led her to a nearby village where an orphaned individual was kept as a pet. The wild population of Sanje crested mangabeys is now thought to number around 2,000–3,000.

MONKEY DISCOVERIES

Among the most remarkable recent events in African primatology was the simultaneous discovery of two populations of a previously undescribed mangabey-like monkey more than 250km apart in southern Tanzania in 2003 – one on the slopes of Mount Rungwe and the other in the remote Ndundulu Forest Reserve in the Udzungwa Mountains. Initially named the highland mangabey (with the original Nyakyusa name preserved in the second part of the scientific binomial *Lophocebus kipunji*), this large and very arboreal grey-brown monkey has a distinctive 'honk-bark' alarm call, an unusually shaggy coat, and a long erect crest on its head. Subsequent to its discovery, DNA tests have demonstrated that this supposed mangabey is actually more closely related to baboons. It is now assigned to the monotypic genus *Rungwecebus*, the first primate discovery of comparable magnitude in 83 years, and is known by the common name of kipunji monkey. The kipunji's range is restricted to three sites with a total extent of less than 75km², and the total population is estimated at a mere 1,100 individuals.

PROSIMIANS

Seldom observed due to their nocturnal habits, the prosimians are often – and somewhat homocentrically – referred to as 'primitive primates' on account of being more closely related to the lemurs of Madagascar than to the diurnal monkeys and apes of the African mainland. Two prosimian forms occur in East Africa – the widespread and diverse bushbabies, discussed below, and the unique potto (*Perodicticus potto*). The latter is a medium-sized sloth-like creature of rainforest interiors (including Kakamega, Kibale, Nyungwe and Bwindi), where it spends the nights foraging upside down from tree branches and can sometimes be located by shining a spotlight into the canopy.

Bushbabies

With their wide round eyes and agile bodies, bushbabies (or galagos) are endearing creatures that can enchant the least sentimental of observers. Two species are widespread in the East African savannah. The greater (or thick-tailed) galago (*Otolemur crassicaudatus*), which weighs up to 1kg, occurs throughout the eastern side of Africa from South Africa to southern Kenya, and produces the eerie and rhythmic screaming call from which the group derives its popular name. The more lightweight northern lesser galago (*Galago senegalensis*), which weighs in at just 150g, is also more often

Lesser galago (SS)

heard than seen, though it is sometimes revealed by tracing a cry to a tree and picking out its eye-shine with a torch. One of the best places to see the thick-tailed greater galago is Shimba Hills Lodge (in the eponymous national park), which is visited by half a dozen every night. The northern lesser galago, with its guileless wide eyes, is often picked up by spotlight in reserves where night drives are permitted.

The greater galago, like most of its kind, takes refuge in a tree hole during the day. (AVZ)

GALAGOS GALORE

In the mid-1970s, six species of galago were recognised by primatologists. Today, that number has risen to more than 20, of which about half occur in East Africa and five are probably endemic to the region. This is due to biologists studying more closely the animals' vocal repertoire, which can provide a more accurate indicator than their appearance of whether or not two populations might interbreed given the opportunity (in other words, whether or not they should be regarded as discrete species). Simon Bearder of the Nocturnal Primate Research Group argues that the implications of these fresh discoveries in galago taxonomy might extend to other 'difficult' groups of closely related animals. He points out that our most important sense is vision, which makes it easiest for us to separate species that rely primarily on the same faculty to recognise or attract partners. It becomes more difficult for us to separate animals that attract mates primarily by sound and scent, more so still if they use senses we do not possess such as ultrasound or electric impulses. 'Such "cryptic" species,' Bearder writes, 'are no less valid than any other, but we are easily misled into thinking of them as being much more similar than would be the case if we had their kind of sensitivity. The easiest way for us to distinguish between free-living species is to concentrate on those aspects of the communication system that the animals themselves use to attract partners.'

East Africa's many large carnivores, such as the spotted hyena, embody nature 'red in tooth and claw'. (AVZ)

CARNIVORES

For most first-time visitors to Africa, the perceived success of a safari rests largely on the quality of carnivore sightings, with big cat encounters being particularly prized. Perhaps this has something to with their resemblance to overgrown pussies, or possibly it's simply because our ancestors expended so much time and effort on keeping out of their way, but there's no doubting the fact that Africa's big cats exude a singular fascination over those who encounter them in the wild. The guaranteed showstopper on any first safari is the lion, but repeat visitors generally place a higher premium on the less readily observed leopard and cheetah. And East Africa's list of carnivores doesn't stop here. A regional checklist of 36 species includes canids such as the jaunty jackals and endangered African wild dog, beautifully marked nocturnal viverrids such as the African civet and the genets, the unfairly maligned hyenas, a host of jittery mongooses, otters and honey badgers, and a trio of secretive small felines.

Carnivores, as the name suggests, are dedicated meat-eaters, with a distinctive tooth pattern designed to process flesh or bones, including lengthened canines, chisel-like incisors, strong molars and a unique pair of meat-shearing premolar teeth known as carnassials. The 280 species recognised worldwide are a diverse bunch, including such varied forms as seals, mongooses and the giant panda (the only strict vegetarian in the order), and ranging in size from the 250g least weasel to the 3,400kg bull elephant seal. Most terrestrial carnivores are solitary, territorial, and approach any encounter with another individual of the same species warily or aggressively, but the order also showcases some of the most complex mammalian social systems, for instance those of the African wild dog, spotted hyena, lion and dwarf mongoose.

Male lion (AVZ)

The ancestors of the modern order Carnivora first appear in the fossil record some 60 million years ago in the form of the miacids, a now extinct family of primitive predators that resembled modern genets. The relationship between different carnivore groups has long been rather controversial among scientists, but currently two distinct suborders are recognised: cats, mongooses, viverrids, hyenas and Madagascan euplerids are placed in the suborder Feliformia (cat-like carnivores); dogs, mustelids, seals, racoons, bears and pandas belong in the Caniformia (dog-like carnivores).

CATS

Stealthy, solitary, secretive and inscrutable, the cats of the family Felidae are the most efficient of mammalian killers, and the most strictly carnivorous, feeding almost exclusively on other warm-blooded creatures, from sparrows and mice to buffalo and giraffe. With their elongated bodies, proportionately small heads, sensitive whiskers, prominent canines and keen bifocal vision, all cats conform to a near-identical body plan, the main physical difference between various species being in their coat pattern and relative size. In evolutionary terms, the past million years might be regarded as the Age of the Felids, with an estimated 60% of modern species having evolved during that short period.

Lion

The lion (*Panthera leo*) is Africa's largest predator and the one animal that everybody hopes to see on safari. With a shoulder height of up to 1.2m and an average weight

Play in lion cubs is a vital part of growing up. (AVZ)

of 150–220kg (though larger individuals have been recorded), it is second in size only to the tiger among the world's 36 cat species. Description is hardly necessary but, for the record, the lion is East Africa's only large tawny to tan-grey cat species. Most adult males sport a unique golden or black mane, while young adults are often lightly spotted – though this is only visible from close up. Lions are the most sociable and least secretive of cats, typically living in prides of five to ten animals and defending a territory of between 20 and 200km², though larger aggregations are a feature of certain parts of East Africa, notably the northern Serengeti, the Maasai Mara and Ruaha National Park. The classic pride structure consists of one adult male, up to three adult females, and an assortment of cubs and juveniles, but male coalitions, generally but not always between litter mates, are regularly noted among larger prides in the Serengeti-Mara. Rivalry between males is intense: battles to take over a pride are frequently fought to the death, with male cubs usually being killed after a successful takeover, while in more stable circumstances a male will boot out his male offspring (or risk having them boot him out) when the youngsters reach an age of around three years. Unsurprisingly, very few male lions ever reach a ripe old age!

The lion's favoured prey is large or medium-sized antelope such as wildebeest, impala and gazelle. They are highly opportunistic, however: lone individuals frequently prey on birds and small mammals; large prides are capable of downing the likes of giraffe and buffalo; and any lion might snaffle up an injured adult or abandoned youngster of any species. Most hunting is undertaken co-operatively by females, with the usual strategy being for one or two females to herd the prey into an 'escape route', where one or more of their allies lie hidden in the grass. In most areas, hunting takes place around dusk or dawn, or at night, but there are exceptions – during the dry season in Nyerere National Park, for instance, lions habitually spend the day lazing close to a water source and picking off any animal that strays within pouncing range on its way to drink. Even when females have done the work, the dominant male will emerge from out of nowhere and exercise his right to feed before the other pride members.

Lions mate every 15–20 minutes while the female is on heat, so if you locate a pair looking a little weary, you won't have to wait long to see a violent coupling. (AVZ)

One of the most extraordinary aspects of leonine behaviour is the mating ritual, which entails a male and female pairing off for several days, copulating at gradually increasing intervals of 12–25 minutes, the act itself often taking less than 60 seconds and being accompanied by considerable snarling and growling, until eventually they are too exhausted and hungry to continue, and return to the main pride.

A termite mound makes a useful vantage point for spotting prey. (AVZ)

When not feeding or fighting, lions are remarkably languorous, with adults spending up to 23 hours of any given day at rest, which means that the anticipation of a sighting is often more exciting than the real thing. The cubs are far more active, however, and will often play and mock fight in a rather kitten-like manner for several hours daily. You're most likely to catch lions in something approaching action mode in the early morning cool (the first 30 minutes after sunrise) and to a lesser extent towards dusk. Lions seldom move any great distance during daylight hours, so if you locate a pride at rest more than, say, 90 minutes after sunrise, odds are it will still be in the same place or very close by the same afternoon, so it's worth checking for activity. By contrast, a pride will almost always cover significant distances at night, the only exception being if it decides to stay close to a fresh kill.

Unusually for cats, lions are almost exclusively terrestrial, and they generally look quite out of their depth on the rare occasions they attempt to climb a tree. There are a few specific areas, however, where tree-climbing behaviour is frequently observed – nowhere more so than the Ishasha sector of Uganda's Queen Elizabeth National Park, but also, to a lesser extent, in Lake Manyara National Park and parts of the Serengeti and Nyerere. The reason for this unusual behaviour is unclear: most likely it started during a biting fly epidemic, and has since been passed on culturally.

Lions naturally occur in any habitat but desert and rainforest, and they ranged across much of Europe and Asia into historic times. The North African and Middle Eastern populations were hunted out in the early 20th century, and the once plentiful Asiatic population has been reduced to about 300 individuals living in one national park in the Indian province of Gujarat. That aside, the global population of wild lions is now all but restricted to the larger conservation areas in sub-Saharan Africa, where they are reasonably common in most savannah and woodland reserves. The Serengeti-Mara-Ngorongoro ecosystem is undoubtedly the best place in the world for lion sightings, but Samburu-Buffalo Springs, Tsavo, Nyerere, Katavi, Ruaha, Queen Elizabeth and Murchison Falls national parks are also very good. Outside of game reserves, however, lion numbers are in rapid decline: according to an assessment undertaken by LionAid (w lionaid.org) in 2020, Africa's total lion population might be as low as 10,000 adults, which represents less than 10% of the 1990 figure.

Leopard

The most widespread and abundant of Africa's large predators, the leopard (*Panthera pardus*) is also the most elusive, and the most ardently sought by seasoned safari-goers. The explanation for this apparent paradox is simple enough: the leopard's resilience in farmland, nature reserves, mountains and periurban forests, where other large predators were long ago hunted out, is due to the same spectral furtiveness that renders it so frustratingly difficult to find on safari. There are many records of individuals living in close proximity to humans for years without being detected, most famously perhaps in Karen Forest on the outskirts of Nairobi, where domestic dogs fall prey to these secretive cats with surprising regularity.

With its spotted coat and powerful pugilistic build, the leopard is the supreme solitary hunter, determinedly nocturnal and so well camouflaged that it will often get within 5m of its intended prey before it pounces. Furthermore, it is unusually versatile in its habitat preferences, as at home in the sun-baked Kalahari sands as in the impenetrable Congolese rainforest. Population densities in East Africa tend to be highest in places that offer plenty of cover, such as riverine woodland and rocky slopes. Indeed, despite considerable persecution over the years, the global leopard population is estimated at around 500,000, concentrated in sub-Saharan Africa, but ranging across the Middle East and Asia as far as Java.

In rainforests and other habitats where lions are absent or irregular, the leopard is typically the apex predator, with males often weighing around 90kg. Elsewhere, leopards typically weigh 60–80kg (females being visibly smaller than males), and habitually store any large kill high in a tree to keep it from hyenas and lions.

With the exception of a female and her cubs, which become independent at the age of two, the leopard is among the most determinedly solitary and territorial of cats, and any chance meeting between two individuals is usually accompanied by real or feigned aggression. A leopard might walk 20km or more in one night, marking and defending its territory against members of the same sex, and while the territories of males and females do sometimes overlap, coupling tends to be an ill-

tempered and abruptly executed business, with the male having no subsequent involvement in rearing the cubs.

The leopard is an adaptable hunter, feeding on anything from medium-sized antelope such as impala or bushbuck, to baboons, hyraxes, hares and birds. You might spot one in pretty much any East African game reserve – but more likely

Left A leopard drinking. (AVZ)

Right Leopards are the most skilled climbers among big cats, and frequently spend the day in a tree. (AVZ)

A leopard generally captures its prey on the ground, using extreme stealth and power. (AVZ)

won't, for there are many places that support healthy leopard populations where you could drive around for months without so much as a nocturnal glimpse. Among the most reliable locations are the Seronera Valley in the south-central Serengeti, where they often rest up in acacia trees during the day, as well as the riverine forest of Samburu-Buffalo Springs, and the scrub lining the Kazinga Channel in Queen Elizabeth National Park.

In Asia, the leopard is also sometimes referred to as the panther, a generic and increasingly archaic term elsewhere applied to the puma and jaguar. The term panther is also sometimes used to denote a melanistic (all-black) leopard, a morph that is quite common in Asia but very rare in Africa, except in a few specific mountain ranges such the Aberdares in central Kenya. Oddly, the name leopard stems from an ancient belief that the familiar spotted version of this big cat represented a hybrid between a lion and a black panther, 'leo' being ancient Greek for lion and 'pardos' for panther.

Cheetah

The cheetah (*Acinonyx jubatus*) is the felid equivalent to the greyhound, with a thrillingly streamlined build and unique semi-retractable claws that assist it in achieving speeds of more than 110km/h in short bursts, outsprinting all other terrestrial animals. Like the leopard, the cheetah has a spotted coat (its name derives from the Hindi *chita*, meaning 'spotted one'), but the two are easy enough to tell apart: the cheetah has simple evenly spaced spots, whereas the leopard's are clustered in rosettes, and the black 'tear marks' running between the cheetah's eyes and mouth are diagnostic. Furthermore, the cheetah and leopard have very different builds (like a whippet and bull mastiff, respectively), and the former is a diurnal hunter that spends its daylight hours pacing restlessly or lying in open grassland, while the latter is very much a creature of shadow, cover and night.

Male cheetahs are generally solitary and highly territorial, though coalitions between two or three brothers are quite commonplace in the Serengeti. Females are less rigidly territorial and related individuals will often have overlapping ranges. Youngsters generally stay with the mother until they reach about two years of age, and for the first few months are easily distinguished from adults by the fluffy white mane down their back. Cheetah cubs suffer a higher mortality rate than other cats, probably as a result of the inherent exposure of the open plains they inhabit, and are unique in that they need to be taught how to hunt.

The cheetah's normal modus operandi when hunting is to creep to within 15–30m of its intended prey, usually a gazelle or other small to medium-sized antelope, before giving chase, then knocking it down and suffocating it with a neck bite. Being less powerful than competitors such as the lion and spotted hyena, a cheetah is often chased off its kill, so it has to gobble down what it can very quickly, consuming up to 10kg of meat in 15 minutes. Unfortunately, the cheetah's hunting habits have been

A female cheetah with cubs. (AVZ)

A cheetah scanning the surrounding plains for prey. (AVZ)

adversely affected in areas where there are high tourist concentrations and off-road driving is permitted – though it has arguably compensated for this in the likes of the Maasai Mara by learning to use parked safari vehicles as cover to stalk its prey.

In evolutionary terms, the cheetah is the last living representative of a felid genus that diverged from the ancestors of other modern cats about 10 million years ago. Several extinct species have been identified from fossils, including the gigantic *Acinonyx pardinensis*, which once ranged throughout Europe and Asia. The low genetic variation and poor adaptability of modern-day cheetahs is thought to be the result of high interbreeding within a bottleneck population that narrowly evaded extinction about 35,000 years ago. Unlike true big cats (*Panthera* spp.), the cheetah is capable of purring but cannot roar. Indeed, the sound it most commonly makes is a peculiar high-pitched twittering that sounds more like a bird or bat than a large carnivore.

The cheetah was widespread in the Middle East and southern Asia into recent historical times, when it was highly prized as a status symbol and pet by the aristocracy, who used it to hunt antelope. Today it is practically endemic to Africa, with fewer than 100 individuals thought to survive in Iran, while occasional unsubstantiated sightings in India and Pakistan are probably a result of confusion with other spotted cats. Even in Africa, recent persecution by humans has led to a massive range retraction and population decline from an estimated 100,000 in 1900 to fewer than 7,000 individuals today, roughly 25% of which are resident in Tanzania and Kenya. The cheetah's main East African stronghold is the Serengeti-Mara ecosystem, where it is common in most open habitats, particularly the southeastern plains of Serengeti National Park into the bordering Ngorongoro Conservation Area, and the Maasai Mara. Elsewhere, it is quite common in Ruaha and Samburu-Buffalo Springs, very rare in Uganda and absent from Rwanda.

Young cheetah cubs have a distinctive silvery mane. (AVZ)

Smaller cats

Three small cat species associated with savannah and woodland are widespread in East Africa, and one localised rainforest species is also present in the region. All are secretive and nocturnal, and seldom seen on safari. The smallest is the African wild cat (*Felis lybica*), which is the direct ancestor of the domestic cat (*F. catus*) and very similar in appearance to a slim, long-legged tabby. The wild cat is an adaptable and skilled hunter of rodents, birds and insects, and it ranks among the most widely distributed of African predators, being absent only from rainforest interiors and true deserts. In some areas, its genetic integrity has been compromised by interbreeding with feral cats, which are occasionally seen in East Africa's game reserves.

The serval (*Leptailurus serval*) is a slender spotted cat associated with moist savannah, swamp and forest-fringe habitats. Its slim, long-legged build superficially resembles that of a cheetah, but it is smaller (shoulder height 55cm), has a proportionately shorter tail (one-

Top African wild cat (AVZ)
Above Serval (AVZ)

third of the body length as opposed to two-thirds) and the black-on-gold spots are elongated into lines towards the head. It also has exceptionally large and prominent ears, which are an adaptation for detecting the movement of prey in long grass. The serval feeds mainly on rodents and birds, which it pounces upon with a spectacular high spring, dispatching its victim with a fatal blow from one of its forepaws. Serval sightings are far from commonplace, with the best chance being in the Serengeti, especially about half an hour after sunrise.

The largest of Africa's 'small cats', the caracal (*Caracal caracal*) strongly resembles the northern hemisphere lynxes with its uniform tan coat and tufted ears. It is a solitary nocturnal hunter, feeding on birds, hyraxes, small antelope and livestock, and ranges throughout the region favouring relatively arid savannah habitats. Similar in appearance but stockier and more golden in colour, the golden cat (*C. auratus*) is a Congolese rainforest species whose range extends into suitable habitats in Uganda, Rwanda and western Kenya. Uncommon and very secretive, it is listed as Vulnerable on the IUCN Red Data List and extremely unlikely to be observed even in the few localities where it is present.

Spotted hyena with young pup at a den. (AVZ)

HYENAS

From Ovid's *Metamorphoses* to Disney's *The Lion King*, popular culture has tended to portray the hyenas of the family Hyenidae in a less than flattering light. The ancients frowned upon their supposed hermaphroditism and believed that an individual hyena could change sex at will, a myth that stemmed from the false scrotum and penis covering the female's vagina. Another ancient myth cited the hyena as a product of interbreeding between domestic dogs and wild wolves, and even today it is widely assumed that the hyenas are a form of canid, though they are actually more closely related to cats and mongooses.

Contemporary legend, meanwhile, portrays hyenas as craven, giggling ninnies whose livelihood depends on scavenging from more noble hunters such as lions and leopards. This negative anthropomorphic perspective that would be daft even if it were true, but is probably false anyway – recent studies in the Serengeti-Ngorongoro suggest that lions scavenge from hyenas just as often as hyenas do from lions. The spotted hyena, in particular, is an adept hunter, capable of killing an animal as large as a wildebeest, though it also derives much of its protein from carcasses left by others.

Three species of hyena occur in East Africa, of which two are very seldom seen: the north African striped hyena (*Hyaena hyaena*), which is pale brown with several dark vertical streaks and an off-black mane, and occurs in dry regions of East Africa; and the shy aardwolf (*Proteles cristatus*), which is closer in size to a jackal and feeds almost entirely on termites. Far more numerous and visible, the spotted hyena (*Crocuta crocuta*) is Africa's second-largest predator, with a shoulder height of 85cm and an average weight of greater than 70kg. It has the powerful build, downward-sloping back, powerful jaws and guileless dog-like expression of all hyenas, but is distinguished from other species by its larger size and blotchy brown coat.

The striped hyena is solitary and reclusive. (AVZ)

Although it is often seen alone, the spotted hyena is a highly sociable animal and can be quite fascinating to observe, living in loosely structured clans of 5–25 animals and performing elaborate dog-like greeting rituals whenever individuals meet. Females, which are stronger and larger than males, lead the clans, and each clan will recognise a strict matriarchal hierarchy headed by the dominant alpha female. The spotted hyena is probably the most common large predator in East Africa. It is most frequently seen at dusk and dawn in the vicinity of game reserve lodges, campsites and refuse dumps, and is likely to be encountered on a daily basis in the Serengeti-Mara-Ngorongoro ecosystem. Even if you don't see it, you are bound to hear it, though its famous 'laugh' is far less commonly heard than a far-carrying haunted 'whoooo-whoop', which rises at the end to form one of the definitive sounds of the African night.

MONGOOSES

Mongooses, which comprise the family Herpestidae, are small carnivores characterised by a slender build, narrow muzzle, long tail, small eyes and ears, non-retractable claws, and uniformly coloured, grizzled coats. Some species are diurnal; others are nocturnal. Based on the behaviour of Asian species, it is popularly assumed that mongooses feed mainly on venomous snakes but this doesn't seem to be the case in Africa, where the favoured prey of most species are rats, insects, molluscs, crabs and other small terrestrial creatures.

Banded mongoose (AVZ)

Mongooses are the most abundant of African carnivores, both numerically and in terms of diversity. Taxonomists now recognise around 35 species, split between 15 genera, all but one of which (the Asian mongooses of the genus *Urva*) are endemic to Africa. You are likely to see several types of mongoose on safari, and many species are also widespread outside of protected areas.

Typical mongooses of the genus *Herpestes*, sometimes referred to as ichneumons, are represented by five species in Africa. Of these, the slender, light grey Egyptian (or large grey) mongoose (*H. ichneumon*) is quite common in East Africa, especially on seasonal floodplains and other grassy habitats close to water. More common, however, is the slender mongoose (*H. sanguineus*) a solitary and diurnal species that is widespread in savannah habitats and suburbia. It has a uniform brown coat and black tip to the tail, which it habitually holds in the air when running.

Also very conspicuous is the banded mongoose (*Mungos mungo*), a ferret-sized dark greyish-brown carnivore, with a dozen or so

Dwarf mongooses often make their home in a termite mound. (AVZ)

faint black stripes running across its back. One of the most sociable mongooses, it is active by day in family bands of around 10–20 members, and might be seen in any savannah or wooded habitat, with the Mweya Lodge in Queen Elizabeth National Park being a particularly good spot thanks to the presence of a habituated group.

Also very common and widespread, the dwarf mongoose (*Helogale parvula*) is a diminutive (shoulder height 7cm), light brown and highly sociable species, often seen around termite mounds, particularly in drier parts of the Rift Valley such as Tarangire National Park. By contrast, the badger-sized white-tailed mongoose (*Ichneumia albicauda*) is the largest species of mongoose in the world. It often appears in the spotlight on night drives, and is a regular visitor to the 'tree hotels' of the central Kenyan highlands. This solitary nocturnal hunter habitually walks with its muzzle to the ground, foraging for the invertebrates on which it mostly feeds. Its bushy white tail is usually diagnostic, though dark-tailed individuals do occur, especially in Uganda.

The marsh mongoose (*Atilax paludinosus*) is another large and solitary species, and has a thick, dark brown coat. It is primarily nocturnal and usually seen close to water, where it is sometimes mistaken for an otter. The rarest mongoose is the Sokoke dog mongoose (*Bdeogale omnivora*), a creamy-grey East African forest endemic, whose confirmed range is confined to the Kenyan coast between Malindi and Mombasa – though it has also long been listed for Tanzania's East Usambara Mountains, based on a 1939 sight record that was substantiated only in 2003.

Semi-habituated genets can be easily observed at certain camps after dark. (AVZ)

GENETS AND CIVETS

Probably the oldest extant carnivore family, Viverridae today comprises approximately 30 species of genet, civet and linsang that range across most of Asia and Africa. Modern viverrids are very similar to fossils of prehistoric species that date back more than 30 million years, and are probably ancestral to most if not all modern carnivore families.

The small-spotted or common genet (*Genetta genetta*) and large-spotted genet (*G. fieldiana*) are the most readily seen viverrids, since both species often become very tame at lodges were they are fed – eg: Ndutu Lodge on the Serengeti/Ngorongoro border, The Ark in the Aberdares, Rubondo Island Lodge and several places in Samburu-Buffalo Springs. More occasionally, they are also seen on game drives – especially night drives, where these are available. Both are very feline in appearance (indeed, they are often but erroneously called genet cats), with a slender, black-spotted torso, a low centre of gravity and a long striped tail. They are best distinguished from each other by the colour of the tail tip: black in the case of the large-spotted and white in the case of the small-spotted.

The African civet (*Civettictis civetta*) is a bulky, long-haired, spotted omnivore that feeds on small animals, carrion and fruit. It is occasionally seen on night drives, walking with a rather deliberate air, nose close to the ground as if following a scent. Several other similar-looking genets and palm civets occur in forested regions, but none is likely to be seen by safari-goers.

DOGS

Probably the most sociable of carnivores, the members of the family Canidae require little introduction, since most wild species are strikingly similar in general appearance and mannerisms to their familiar domestic kin. Possibly as a result of this strong resemblance, the five dog species that inhabit East Africa seldom register high on safari 'wish lists', and you'll often see tourist vehicles drive straight past a family of jackals (the most common wild dogs) as if they weren't there. In terms of behaviour and interaction, however, these small and rather jaunty-looking canids can be fascinating to observe, especially when denning, as can the insectivorous bat-eared fox. The largest and most endangered East African canid, placed in a monospecific genus, the somewhat prosaically named African wild dog is eagerly sought by experienced safari-goers as a result of its growing scarcity.

African wild dog

The African wild dog (*Lycaon pictus*), also known as the African hunting dog or painted dog, is the largest East African canid, distinguishable from all other species by its ultra-sociable habits and cryptic black, brown and cream coat. Typically living in packs of 5–50 animals, it is a ferocious co-operative killer, with several individuals literally tearing apart their prey on the run.

The wild dog was once so common that it was treated as vermin in much of Africa, but it has suffered enormous losses over the past few decades, partially through direct persecution, and partially through susceptibility to

Wild dogs have distinctive ears. (AVZ)

diseases spread by domestic or feral dogs. It is now the second most endangered of Africa's large carnivores (after the Ethiopian wolf), with the total wild population estimated to stand at around 6,000, a 50% increase since the turn of the millennium. In many reserves where wild dogs were formerly common, they are now all but extinct, including the Serengeti-Mara-Ngorongoro ecosystem, where Jane Goodall and Hugo Von Lawick undertook their famous behavioural study of wild dogs in the 1970s.

In this context, the importance of Nyerere National Park to wild dog conservation is difficult to overstate. Recent surveys indicate the reserve supports a total population of more than 1,000 individuals, which is almost twice as many as any other African country, let alone any other game reserve. Elsewhere, Ruaha and Mikumi national parks, also in southern Tanzania, have good reputations for wild dog sightings. The only place in Kenya where wild dogs can be reliably observed is the Laikipia Plateau.

African wild dogs greeting pack members. (AVZ)

The black-backed jackal is an alert and versatile predator. (AVZ)

Jackals and wolves

Far more common than African wild dogs but no less fascinating to watch are the region's three jackal species, which each stand about 40cm at the shoulder. Most widespread is the black-backed jackal (*Lupulella mesomelas*), an opportunistic feeder that is most often seen singly or in pairs at dusk or dawn, and has an ochre coat with a prominent black saddle flecked by a varying amount of white or gold. It is probably the most frequently observed small predator in East Africa, and its eerie call is a characteristic sound of the bush at night.

The similar but less common side-striped jackal (*L. adusta*) is more cryptically coloured, with an indistinct pale vertical stripe on each flank and a white-tipped tail, and is distributed throughout most of East Africa, though it is most likely to be seen in southern Tanzania. The African golden wolf (*Canis lupaster*) is a cryptic northern hemisphere species that was long thought to be a subspecies of Eurasian golden jackal. However, DNA tests undertaken in 2015 indicate it is a distinct species more closely related to the grey wolf and coyote to any jackal. Its range extends southward to the Serengeti and Ngorongoro Crater, where it lives alongside both true jackal species.

Side-striped jackal (AVZ)

Jackals form stable pairs, though larger aggregations occasionally gather at a kill. A particularly good time to watch jackal interaction is during the denning season, towards the end of the rains, when the (often very inquisitive) pups will play openly outside den entrances close to the road, and the parents regularly return to bring them food. An unusual feature of jackal family life is that pup-rearing duties are often shared by 'helpers', the name given to sub-adult or young adult offspring who bring additional food to the den, as well as babysitting the pups when the parents are out foraging.

Golden jackal with pup. (AVZ)

Bat-eared fox

The bat-eared fox (*Otocyon megalotis*) is smaller than a jackal, and easily identified by its combination of uniform fawn to silver-grey coat, huge ears and black eye-mask. Primarily an insectivore, it tends to be more nocturnal at hotter times of year and more diurnal in the cooler months, but pairs and small family groups might be seen resting up in the shade at any time of the day. Associated with arid climates and open country, the bat-eared fox is quite common in the Serengeti-Mara ecosystem and likely to be encountered at least once in the course of a few days' safari in this area, particularly during the denning season (November and December).

Bat-eared foxes are easily located during the denning season, when they tend to be most sedentary. (AVZ)

Honey badger (AVZ)

WEASEL FAMILY

Mustelidae, the weasel family, is the most diverse of the carnivore families, with an estimated 55 species worldwide. Six species occur in East Africa, including the striped polecat or zorilla (*Ictonyx striatus*), a common but rarely seen nocturnal creature with black underparts and bushy white back, and the similar but much more scarce African striped weasel (*Poecilogale albinucha*).

The Cape clawless otter (*Lutra capensis*) is a large semi-aquatic mustelid, with a streamlined body, and brown coat with a white collar. It is common in suitable freshwater habitats throughout most of the region, but is replaced by the similar Congo clawless otter (*L. congica*) in the west. You are more likely to see the smaller spotted-necked otter (*Hydrictis maculicollis*), which is darker with white spots on its throat, and is often observed in Uganda's Lake Bunyonyi as well as on Rubondo Island in the Tanzanian waters of Lake Victoria.

Arguably the most interesting of the region's mustelids is the honey badger (*Mellivora capensis*), a compact and famously aggressive street-fighter that would probably rank as the most terrifying African carnivore were it not for its lightweight (up to 12kg) status. Black-bodied with a striking grey-to-white cape down its back, a deceptively puppyish face, heavy bear-like claws and a blunt muzzle, the honey badger is also known as the ratel. It is rumoured to have a symbiotic relationship

with the greater honeyguide (*Indicator indicator*). This bird is said to use its call to lead the ratel to a bees' nest or hive, wait for it to tear the nest open, then feed on the scraps it leaves behind. As well as honeycomb, its opportunistic diet includes insects, rodents, birds, fish, snakes, scorpions, fruit and the soft parts of tortoises and terrapins. It is even alleged to run beneath male buffalo and bite off the poor creatures' testicles! The honey badger is among the most widespread of African carnivores, but it is thinly distributed and – being primarily nocturnal – rarely seen.

The Cape clawless otter is the larger of two otter species that inhabit freshwater habitats in East Africa. (JC/D)

The large tusks displayed by these bulls in Ngorongoro Crater are now a rarity, thanks to commercial poaching during the 1980s. (AVZ)

ELEPHANTS

Intelligent, highly sociable and playful, elephants are among the most entertaining animals to observe for extended periods. They are also perhaps the most physically intimidating of living land creatures, on account of their immense bulk, fierce trumpeting call and an unpredictable temperament that occasionally results in tragedy. Like humans, elephants are among the few mammals capable of significantly modifying their environment – especially when concentrated populations are forced to live in restricted conservation areas, where their habit of uprooting trees with tusks and trunk can cause serious deforestation and environmental degradation.

ELEPHANT ADAPTATIONS

The elephant is unique among living mammals in possessing a trunk, which it uses in a variety of ways – to reach into high branches for leaves, to shake a tree trunk to dislodge ripe fruit, to tear up food before placing it in the mouth, or to suck up water for drinking or to spray on its back. An elephant's trunk contains several thousand small individual muscles, giving it immense strength yet also making it sufficiently sensitive to isolate and tear out a single blade of grass. When two familiar elephants meet, they greet each other by intertwining their trunks. This versatile appendage is also used for play wrestling, during courtship, and in displays of dominance (trunk raised) or submission (trunk down).

Another adaptation unique to the elephant is its outsized tusks, an extension of the second incisors that grow continuously throughout its life. The longest recorded pair of tusks originated from the DRC and measured 3.35m and 3.49m, while

Young elephants on a river bank. (*left*, AVZ)

On warm days, elephants often follow a drinking session with a playful splash. (*below*, AVZ)

When no surface water is available, elephants will dig a well in a dry riverbed. (AVZ)

the heaviest single tusk on record weighed 117kg. Like the trunk, the tusks are highly versatile tools, used among other things to dig for salt and water, to tear the bark off trees and pulp the wood inside, to clear a path of obstructions, and more occasionally in defence or in conflicts between rivals. Most elephants are either right- or left-tusked, with the master tusk usually being shorter and having a more rounded tip as a result of greater wear and tear.

Elephant dentition is also unusual in that – tusks aside – the teeth are replaced five times during its lifetime. The last set of teeth usually starts to wear down when the animal reaches its early 60s, forcing it to seek out relatively soft food such as wet grasses and aquatic vegetation. In the wild, most elephants die a cruel, lingering death from starvation once their last teeth fall out or are unable to process enough food. It may well be that the tendency among elderly elephants to forage in marshy areas lies behind legends of elephant graveyards.

It was long thought that aural communication between elephants was limited to the occasional bout of trumpeting, but it has recently been discovered that they can communicate using low-frequency (infrasound) rumblings that travel through the earth and are picked up by the skin on the trunk and feet, allowing dispersed herds to co-ordinate their movements over a vast area. Elephants also have a good sense of smell: when faced with an unfamiliar situation or potential threat, members of a herd will often raise their trunks (which contain the olfactory organs) and move them around slowly to investigate further, looking rather like a gang of pythons entranced by a snake charmer.

As mixed grazer-browsers, elephants have a rather inefficient digestive system, and although they spend 12–15 hours of the day eating grass and other vegetable matter, more than 50% of the daily intake of around 200kg is defecated without having been digested. An elephant also drinks up to 200 litres daily, typically arriving at a water source 3–4 hours after sunrise. On a hot day, a herd might linger until late afternoon to wallow or spray themselves with water, which helps prevent the skin from burning and dehydrating. Also important to temperature regulation, the large thin-skinned ears are flapped incessantly in hot weather to create a local breeze that cools the circulating blood through a dense network of subcutaneous blood vessels.

ELEPHANT ORIGINS

Elephantidae is the only extant family in the order Proboscidea (animals with trunks), which evolved from the same semi-aquatic ancestor as Sirenia (sea-cows) and Hyracoidea (hyraxes). The fossil record indicates that more than 350 proboscid species have walked the earth at some point, inhabiting every continent except for Antarctica and Australia. Today, three living species are widely recognised: Indian elephant (*Elephas maximus*), African bush elephant (*Loxodonta africana*) and African forest elephant (*L. cyclotis*). The African bush elephant, the world's largest terrestrial mammal, is much bigger than either of the other species and common in many parts of East Africa. The smaller, hairier and straighter-tusked African forest elephant is a west/central African species whose range nudges into western Uganda, where it is most likely to be seen in Toro-Semliki Wildlife Reserve.

In common with rhinos, modern elephants are relicts of a family that once embraced numerous genera and species. Unlike rhinos, however, elephant diversity peaked as recently as a million years ago, and it would be far greater today were it not for human persecution, which was partially or wholly responsible for the extinction of as many as a dozen taxa, including the woolly mammoths that survived on the Arctic island of Wrengal until 1500BC, the hog-sized dwarf elephants that inhabited several Mediterranean islands until a few thousand years back, and the north African subspecies of *L. africana* on which Hannibal crossed the Alps.

Intriguingly, it has been theorised that human activity is responsible for moulding the African bush elephant, which suddenly appears on the fossil record about 20,000 years ago, replacing the *Elephas* species that had dominated the African savannah for the previous 4 million years. The theory is that *Elephas* was hunted to extinction, allowing the forest elephant to expand its range into the savannah and to split into two subspecies. A more recent and verifiable incidence of selective breeding influenced by human activity – in this case ivory hunting – has been a continent-wide increase in the incidence of tusklessness, a hereditary genetic abnormality that now afflicts up to 30% of the population in some areas, as opposed to an average of 1% in 1930.

A female elephant suckles her calf for up to four years. (AVZ)

ELEPHANTS TODAY

The common species in East Africa is the bush elephant, though the range of the forest elephant does extend into certain forests in western Uganda. This is the largest living land animal, typically weighing around 6,000kg. The largest individual on record, now mounted and on display in the Smithsonian Museum, clocked in at an immense 12,000kg. African elephants live in close-knit matriarchal clans, in which the eldest female is generally dominant over her sisters, daughters and granddaughters. Males generally leave the family group at around 12 years to roam singly or form bachelor herds.

The African elephant also has the longest gestation period of any land animal (21–22 months), with females typically becoming pregnant for the first time in their early teens and giving birth at intervals of around five years until they reach their late 50s. With calves typically weighing around 100kg at birth and labour lasting for more than 10 hours, twins are – understandably – a rarity, and larger litters are unheard of. Unlike animals that produce offspring more regularly or in greater volumes, each baby elephant represents a major genetic investment to the herd and it is raised communally, with the mother and the other adult females maintaining a vigilant watch until it grows large enough to be safe from predators – something it does at a rate of around 1kg daily.

Elephants are relatively widespread and common in East Africa's major reserves, but they are increasingly scarce outside protected areas, the result of centuries of pressure from ivory and, to a lesser extent, trophy hunters. Combined with the habitat destruction associated with Africa's burgeoning human population, this led the number

Mature bull African bush elephant, Ngorongoro Crater (AVZ).

of elephants to plummet from around 5–10 million continent-wide in the early 20th century to perhaps 1.3 million in the late 1970s. The 1980s was a particularly torrid time in East Africa, with ivory poachers killing up to 90% of the elephant population in former strongholds such as Tsavo, Nyerere and Murchison Falls national parks, leaving the total continental population at an all-time low of 350–500,000 in the early 1990s.

A CITES ban on the ivory trade assisted East Africa in wresting control from the poachers during the course of the 1990s and early 21st century, during which time the African bush elephant population of most major reserves stabilised or increased. Today, it probably stands at around 500,000, roughly half of which is split between Botswana, Zimbabwe and South Africa, where numbers are generally on the increase. Kenya's elephant population increased from 16,000 to more than 35,000 since the early 1990s and Uganda's from fewer than 1,000 to around 8,000 (with more than 4,700 counted in Queen Elizabeth National Park alone in a 2023 aerial census). As recently as 2009, Tanzania was the world's most important elephant stronghold, boasting a total population of almost 115,000 in 2009, up from a 1989 nadir of around 50,000. However, a renewed bout of commercial poaching caused the Tanzania elephant population to plummet more than 60% in five years to an all-time low of 43,000 in 2014. It has since increased to around 60,000. The Rwandan population, restricted to Akagera National Park, amounts to around 140 individuals, a 200% increase since the turn of the millennium. In East Africa, the top places to see giant tuskers and elephant interaction up close are probably Amboseli National Park and the Ngorongoro Crater floor.

In practical terms, the elephant is probably the most dangerous animal to safari-goers, if only because it is large and strong enough to tear open a car and injure or even kill the passengers. Fortunately, however, an agitated elephant will almost invariably mock charge and indulge in some hair-raising trumpeting before it attacks in earnest, and it will seldom take further notice of a person or vehicle that backs off at this display of unease. As a rule, elephants are most edgy around vehicles in areas when they have recently experienced poaching or that border hunting concessions (for instance Ruaha, Nyerere and Murchison Falls), while they tend to be very relaxed in areas where people aren't perceived as a threat.

Almost any male elephant could attack unprovoked when it enters musth, a periodic condition whose name derives from a Hindi word meaning madness, and which is recognisable by the thick secretion that is discharged through the temporal ducts to the sides of the head. Although the exact nature of musth is unclear, it is evidently linked to dominance and sexual behaviour, and is characterised by a 50-fold increase in testosterone levels – which in turn leads to overexcitement, aggression and unpredictability. Likewise, a female with a young calf can also be aggressive. For this reason, it's advisable to keep the car engine running in the vicinity of an elephant until you are certain it is relaxed, and to avoid allowing your car to be boxed in between an elephant and another vehicle. If an elephant does threaten a vehicle in earnest and backing off isn't an option, then revving the engine hard will generally dissuade it from pursuing the contest. Needless to say, it's also wise to give a wide berth to any elephants you see lurking around camp or on a foot safari.

Portrait of a plains zebra. (AVZ)

ODD-TOED UNGULATES

The term ungulate (literally 'hoofed animal') describes a varied selection of herbivores, all of which either walk on hoofed toes or, in the case of whales and manatees, descend from ancestors that did. Two orders of true ungulate are recognised. Rhinos, horses and tapirs, the last of these found only in South America and southeast Asia, are collectively classified as odd-toed ungulates (order Perissodactyla). They can be distinguished from even-toed ungulates (order Artiodactyla) by their toe pattern, with the third digit being the most prominent in all species. In their heyday, some 50–20 million years ago, Perissodactyla species were the most important herbivores on earth, comprising at least a dozen families, many of which sound like the stuff of science fiction. The chalicotheres, for instance, had elongated forelimbs and propelled themselves in a similar manner to gorillas, while *Paraceratherium*, a genus of hornless rhino, may well have been the largest terrestrial mammals ever, standing more than 5m high, almost 10m long, and weighing 15–20 tonnes.

The decline of the perissodactyls was linked to the emergence of grasses, a relatively coarse food source that became pre-eminent about 20 million years ago and was more easily processed by the complex stomachs of even-toed ruminants. A limited number of species thrived until about 10,000 years ago, after which refinements in human hunting techniques added to the existing pressure created by habitat change, a process that accelerated wildly over the past century. Shockingly, of the world's 17 extant perissodactyl species, two survive in a domesticated form only, while the IUCN lists one as Extinct in the Wild, three as Critically Endangered, seven as Endangered and four as Vulnerable. That leaves the plains zebra as the world's only non-threatened wild perissodactyl, a fact that places a fresh perspective on the sight of a herd of these handsome and unique survivors galloping across the African plains.

ZEBRAS

Among the most familiar and popular of African safari creatures, zebras are essentially striped horses, and are placed in the same genus *Equus* as their more monochrome domestic counterparts. Although they belong to a very ancient order, modern zebras probably evolved within the past 2 million years, which is why they compete successfully with antelope and other herbivores in the grasslands and savannah of East Africa. Each individual has a unique stripe pattern, just as with

A dust-kicking herd of plains zebras (*Equus quagga*) drinking in the Seronera River as it flows through Serengeti National Park, Tanzania. (AVZ)

human fingerprints. The purpose of these stripes has long been a subject of debate. The most widely propagated theory is that they act as camouflage, breaking up the animal's body shape as it moves through the long grass, a theory that fails to explain the benefit of stripes to zebras living in arid or short-grass plains. Another theory is that the striping is visually confusing to predators when the herd scatters upon being chased. It is also possible that it serves some unknown sexual role, or is simply a residual feature retained from ancestral stock.

Plains zebra

Two species of striped equid occur in East Africa, the most widespread and common being the plains zebra (*Equus quagga*) – often referred to as Burchell's zebra, a designation more properly reserved for one specific southern African subspecies (*E. q. burchelli*). This stocky grazer stands up to 140cm high and weighs around 250–300kg. With a natural distribution ranging from southern Ethiopia to South Africa, it still occurs in most national parks and other savannah reserves in eastern and southern Africa, where it is often seen in large numbers alongside various antelope, most often wildebeest, gazelle, impala and hartebeest. Large aggregations are generally rather ephemeral, related to food or water sources, or to local migrations. The core social unit is a small, stable and aggressively defended but non-territorial herd, which consists of a stallion with exclusive breeding access to a harem of up to five mares, who usually each produce a foal every one or two years during the rainy season. The associated offspring stay with the parental herd until they reach adulthood, when they typically form another herd, or migrate to one. The plains zebra is not endangered (though the nominate subspecies, a South African endemic, was hunted to extinction in the 19th century), and is seen in large numbers in most Tanzanian and Kenyan savannah reserves, with densities probably being highest on the Serengeti-Mara-Ngorongoro ecosystem. Largely as a result of hunting, its modern Ugandan range is limited to Lake Mburo and Kidepo national parks, while small numbers still persist in Rwanda's Akagera National Park. Visitors from southern Africa might notice that East African plains zebras lack the shadow stripes associated with their southern counterparts.

Grevy's zebra

The dry country equivalent to the plains zebra is the quite magnificent Grevy's zebra (*E. grevyi*), which is significantly larger (shoulder height 150cm, weight up to 430kg) and can be distinguished by its much narrower striping, white belly and large, almost bear-like round ears. This species is far more territorial and less herd-oriented than the plains zebra, presumably as an adaptation to the paucity of food and water in arid environments. A stallion will stake out a territory of up to 10km^2 and will generally enjoy mating rights with any oestral female that passes through – though it appears that he will tolerate other males within the territory provided that they don't infringe on those rights. Listed as Endangered, Grevy's zebra was once widespread in the Horn of Africa, but it is now thought to be extinct in three countries (Djibouti, Eritrea and Somalia) and on the verge in southern Ethiopia, making it virtually endemic to northern Kenya. Here the total population of fewer than 2,500 is concentrated in the private Lewa Downs Conservancy and Samburu-Buffalo Springs, where it is easily observed.

Grevy's zebra of northern Kenya has a more densely striped coat than other species. (AVZ)

Black rhino with oxpecker. (AVZ)

RHINOCEROSES

Their tank-like build, armoured hide and fearsome horns notwithstanding, rhinos are the scarcest of the so-called 'Big Five' in East Africa, largely as a result of poaching. The two African species are commonly referred to as the white and black rhino, though both are similarly grey in coloration. An oft-repeated safari legend has it that the misnomer 'white' derives from the Afrikaans *weit lippe* (wide lips), a reference to the flat mouth that enables the white rhino to mow grass so efficiently. A more likely derivation, perhaps, is the slightly archaic *weid* (meadow or pasture), which describes the grassland on which these wide-lipped creatures graze. Whatever the truth of the matter, the easiest way of distinguishing these otherwise similar species is mouth shape, which is far narrower in the case of the black rhino, with a distinctive hooked upper lip designed for browsing rather than grazing. The black rhino occurs naturally in all four East Africa countries, but is extinct in Uganda. Within East Africa, the white rhino was naturally restricted to northwest Uganda, where it is extinct, but the southern subspecies has been introduced to Kenya, Uganda and Rwanda.

The world's five surviving rhino species, with their combined global population of fewer than 30,000 individuals, are possibly the closest thing we have to living fossils among the larger ungulates. They are the last representatives of a once prolific ungulate group represented by more than 25 fossil genera, ranging from the swift-moving dog-sized *Hyracodon* to the ponderous gargantuan *Paraceratherium*, which were taller than giraffes and three times heavier than the largest elephant. These gigantic forms survived until relatively recent times. The woolly rhino (*Coelodonta antiquitatis*) which had thick mammoth-like fur and a flattened horn enabling it to push aside snow, inhabited the northern steppes of Eurasia until about 8,000 years ago, and is depicted in several cave paintings from that era. Some experts also believe that the 6m-long giant unicorn rhino (*Elasmotherium sibiricum*), which sported a horn as tall as a person and had a rather horse-like gait, survived into historic times. This monster of a rhino may be the substance behind stories of a gigantic, single-horned black bull in certain Russian folklores, as well as being the source of the unicorn legend.

THE DILEMMA OF A HORN

The demise of rhinos is primarily a function of climatic/environmental change and competition with other herbivores, but over-hunting by humans hastened the extinction of the woolly rhino, and is directly responsible for more recent population declines that have placed all five living species on the Critically Endangered list at some point in the past century. In 1892, for instance, the first European to visit the Ngorongoro Crater remarked on the large numbers of rhino there, and promptly downed seven of the unfortunate beasts to prove his point, while in the mid 1940s the aptly named J A Hunter, by appointment of the colonial government of Kenya, shot more than 1,000 rhinos in one year in the vicinity of Tsavo National Park. Even so, the African rhino population was estimated at 75,000–100,000 in the early 1970s, when the creatures were plentiful in the likes of Nyerere (3,000 individuals), Tsavo (5,000) and Murchison Falls (400).

This situation proved to be short-lived, following an outbreak of commercial poaching for rhino horn, which is made of compressed hair and valued as a medicine and aphrodisiac in parts of Asia, and as a dagger handle in Yemen, where a horn could fetch as much as US$400,000 per kilogram on the black market. In Asia, the wild population of Indian rhino stands at 4,000, a significant increase since the turn of the millennium, but the Sumatran rhino is now listed as Critically Endangered, with fewer than 50 individuals surviving, while the Javan rhino is doing only slightly better with a population of around 75. All things being relative, the situation in Africa is somewhat less critical, with an estimated 6,500 black and 20,000 white rhino still thought to be living in wild or semi-wild conditions, but these populations are concentrated in southern Africa, where rhino poaching has increased greatly since the early 2000s, though the birth rate still exceeds the mortality rate. The total rhino population for East Africa hit a nadir of fewer than 500 individuals in the late 1980s, though it has recovered to around the 1,500 mark today.

Black rhinos are notable for their 'saddle back' and hooked (as opposed to square) lip. (AVZ)

Black rhinoceros

The black (or hook-lipped) rhinoceros (*Diceros bicornis*) is significantly the smaller of the two African species, but unless you are able to compare the two side by side, it is more easily distinguished by its hooked upper lip, saddle back, lack of a shoulder hump, and habit of holding its relatively small head high. In southern Africa, where it competes with the grass-eating white rhino, the black rhino is associated almost exclusively with dense woodland, but in East Africa it often forages in more open grassy habitats. This tolerance of habitat doesn't extend to human beings: the black rhino is legendarily mean-tempered and might charge at any provocation, in which case the best recourse for any pedestrian in its line of fire is to climb the closest tree. These days, any black rhino sighting should be considered a stroke of luck. The best place to seek one out is the Ngorongoro Crater, where a resident population of 10–20 animals is often seen emerging from the Lerai Forest shortly after dawn to hang about on the open plains in the middle of the day. In Kenya, the black rhino is a reasonably regular nocturnal visitor to The Ark (a tree hotel in the Aberdares) and there are decent populations in Lake Nakuru, Meru, Tsavo, the Maasai Mara and the privately protected Laikipia Conservancy. Unfortunately it has been poached to extinction in Uganda, where the last sighting occurred in Murchison Falls in 1983. It was reintroduced to Rwanda's Akagera National Park in 2017.

White rhinoceros

Weighing up to 3,600kg, the white or square-lipped rhinoceros (*Ceratotherium simum*) is the heaviest living terrestrial animal after the elephants, but is generally regarded as a more peaceable creature than its hook-lipped cousin. Prehistoric rock art and skeletal remains in the Sahara suggest that this species once ranged northward to the Mediterranean, but its historical range consists of two geographically separated subspecies, with the northern white rhino (*C. s. cottoni*) being centred on the savannah belt between the Congo Basin and the Sahel, and the southern white rhino

(*C. s. simum*) occurring south of the Zambezi. This is the only rhino species not IUCN listed as Endangered (or worse), with a wild population estimated at up to 20,000 animals. This population is concentrated on South Africa, where the southern race made a remarkable comeback from the brink of extinction, with every last individual on Earth today being descended from a bottleneck population of around 15 individuals in the Imfolozi-Hluhluwe Game Reserve in the 1930s.

The northern race of white rhino was still quite common in Murchison Falls National Park, but the last known Ugandan representative was shot in 1982. Its last refuge in the wild was Gambaga National Park in the war-torn DRC, where a 2006 aerial census counted just two individuals. It was declared extinct in 2011, though two partnerless females remain alive in captivity. More of a sop to tourism than of any serious conservation value, four southern white rhinos were introduced to Uganda in 2002, and initially caged at the Entebbe Wildlife Orphanage, but have since been relocated to the private Ziwa Rhino and Wildlife Ranch, where the population now stands at more than 40 individuals and guided rhino-tracking excursions are offered. The northern white rhino doesn't occur naturally in Kenya but the descendants of introduced southern white rhinos are present in the likes of the Laikipia Plateau and Lake Nakuru National Park.

A female white rhino and calf, with the adult clearly showing the wide lips from that distinguish it from the hook-lipped black rhino. (AVZ)

The warthog is the commonest African swine and is likely to be seen in most savannah reserves. (AVZ)

EVEN-TOED UNGULATES

The even-toed ungulates (Artiodactyla) are the most numerous large mammals, with some 220 species occurring worldwide. Along with dogs and cats, they are also the most closely associated with humans, with farmyard animals such as cows, sheep and pigs being among the order's better-known (and better-tasting) representatives. In Africa they make up a far higher proportion of the wild (and domesticated) herbivore biomass than any other mammalian order.

Some artiodactyls have a spreading foot with four digits, but most have a cloven hoof with just two. The ten families are traditionally split between three suborders: Suiformes (hippos, pigs and peccaries), Tylopoda (camels and llamas) and Ruminantia (giraffes, deer and bovids). However, molecular studies have established that cetaceans (whales and dolphins) are artiodactyls (also referred to by some as Cetartiodactyla), which means, for instance, that hippos are actually more closely related to whales than they are to pigs. Artiodactyls first appear on the fossil record 50–60 million years ago. They initially inhabited marginal niches, where their complex digestive systems enabled them to extract maximum nutrition from low-grade fodder. The order's rise to prominence started about 20 million years ago, and was linked to the global spread of grasses, which are tough to digest and ill-suited to more rudimentary perissodactyl (odd-toed ungulate) stomachs. The cud-chewing ruminants are the most 'advanced' artiodactyls in this respect, having four-chambered stomachs that process vegetation other animals would find indigestible. Today, ruminants are comfortably the most successful herbivores on earth, and comprise almost 200 species of deer, chevrotain, giraffe, cattle, sheep, antelope and goat worldwide.

WILD PIGS

The best-known member of the family Suidae is the farmyard pig (*Sus domesticus*), but some 15 wild species are also recognised around the world. Most are somewhat hairier than their domestic cousins, but otherwise conform to a typical piggy shape, having a barrel-like torso, short legs, a large wedged head and small out-turned tusks. Famously unfussy diners, wild pigs eat anything from fruits and crops to carrion and newborn animals, but are particularly partial to roots and bulbs, which they dig up with their tusks and elongated snout. All the African species are sociable, travelling in family parties comprising an adult male and female, and up to five sub-adult offspring. Most have a strong denning instinct, nesting in burrows that they dig themselves or appropriate from the likes of aardvarks. Wild pig diversity in sub-Saharan Africa peaked about 5 million years ago and has since declined to five endemic species split between three genera.

Warthogs

The common warthog (*Phacochoreus africanus*) is an unusually slender and long-legged swine that stands up to 80cm high at the shoulder, and is named for the trio of callus-like 'warts' on its face. It has a grey, near-hairless body, topped with a long mane from its neck to its rump, and its impressive upward-curving tusks are larger than those of other African swine. Warthogs are most often seen in family groups,

which tend to trot away briskly with an air of determined nonchalance, their long tails raised stiffly in the air.

In most East African savannah reserves, the common warthog is the only wild pig you are likely to see in daylight. In the little-visited far northeast of Kenya, it is replaced by the uncommon desert warthog (*P. aethiopicus*), a very similar looking (but dentally distinct) species, whose range is otherwise confined to arid parts of Somalia and possibly Ethiopia. The best sites for seeing desert warthog are Marsabit and Malka Mari national parks, and Sibiloi National Reserve.

Bushpig

Bulkier, hairier and more warmly coloured than the warthogs, the bushpig (*Potamochoerus larvatus*) is also very widespread in East Africa, but hard to see due to its nocturnal

The normally secretive bushpig is frequently seen after dark from the tree hotels of the central Kenyan highlands. (AVZ)

habits, secretive nature and preference for dense vegetation. Bushpigs occupy a variety of wooded niches, from montane forest interiors to strips of riparian woodland running through otherwise dry savannah. Individuals vary greatly in colour, ranging from chestnut red to greyish brown, and most have a pale grey or white crest along the back. The legs are shorter and the body rounder than those of the warthog, from which it is easily distinguished by its more hirsute appearance and smaller (largely concealed) tusks. You might stumble across a bushpig almost anywhere in East Africa, but the best places for regular sightings are The Ark and Treetops in the central Kenyan highlands. In the far west of Uganda, the bushpig may exist alongside the red river hog (*P. porcus*), a striking red, long-eared pig of the west-central African rainforest.

Giant forest hog

The giant forest hog (*Hylochoerus meinertzhageni*) can reach up to 2m long and 250kg in weight, making it the world's heaviest pig. Yet despite its monstrous proportions, it eluded Western science until 1904, when one was shot in the then Belgian Congo. This species is the hairiest African pig, and dark grey-brown, with a naked face, wide snout and fairly large tusks. Nocturnal and strictly vegetarian, it is typically associated with lowland rainforests, though the eastern race also occurs in montane forests on Mount Kenya, where it is a regular nocturnal visitor to Mountain Lodge. The best place to see giant forest hogs by day is Channel Drive in Queen Elizabeth National Park, where a small thicket-dwelling population has become unusually habituated to cars and is quite often seen crossing to drink from the Kazinga Channel.

The giant forest hog is the world's largest member of the pig family. (AVZ)

HIPPOPOTAMUS

The gregarious hippopotamus (*Hippopotamus amphibius*) is perhaps the most characteristic resident of East Africa's rivers and lakes – and indeed its name, derived from the Greek *hippos potamos*, literally means 'horse of the river'. Typically seen in groups of around 15 grunting, yawning individuals, this massive animal is well adapted to an aquatic existence, with ears, eyes and nostrils placed high on the roof of the skull, allowing it to spend most of the day almost entirely submerged. Adults typically resurface to breathe every 5–8 minutes, and are able to do so while sleeping under water. Remarkably, baby hippos are born under water and must swim to the surface to take their first breath.

Despite its aquatic lifestyle, the hippo feeds almost entirely on land, typically emerging from the water at dusk or by night to crop some 45–60kg of grass daily, and often wandering far from water to find sufficient grazing. Contrary to what you might imagine, it is also a poor swimmer, spending most of its time standing or lying

Despite its monstrous canines, the hippo is a strict vegetarian. (AVZ)

in suitably shallow water, where the same group is generally seen in the same place day after day, relocating only when forced to do so by fluctuations in the water level. As a result, hippos are strongly territorial, typically living in herds (sometimes called pods or bloats) of up to 30 animals – occasionally more – over which a dominant male presides, defending his patriarchy to the death.

The hippo is unmistakable, with its purplish-grey hairless hide, pinkish undersides and cheeks, barrel-like torso and stumpy legs, not to mention the enormous mouth that opens regularly in a 'yawn' (actually a threat display) to reveal a set of positively monstrous canines. It vies with the white rhino as Africa's second-largest land animal, averaging more than 3m in length and weighing up to 3,200kg – though 4,000kg specimens have been recorded.

Despite its immense bulk and aquatic habits, the hippo is impressively mobile on land, attaining a speed of more than 35km/h when disturbed, when it typically heads straight for the safety of the water and mows down anything in the way, making it

A female hippo is highly protective of her calf. (AVZ)

a serious hazard to humans. It is also a long-lived animal, often attaining an age of 40-plus in the wild and passing the half-century mark in captivity.

The family Hippopotamidae is now endemic to sub-Saharan Africa, the only other living species being the pygmy hippopotamus, a localised west African forest endemic. However, the common hippo was resident on the Egyptian Nile into historic times (as alluded to by Pliny the Elder), while three other species have recently become extinct on Madagascar, one within the past 1,500 years, and a dwarf Cypriot species became extinct at the end of the Pleistocene, possibly as a result of human intervention.

The hippo's African range has contracted greatly in the past century and continues to do so, but it remains quite common in most suitable East African habitats, particularly in protected areas. The ultimate hippo destination in East Africa is Katavi in western Tanzania, where aggregations of several hundred animals gather in the late dry season (September–November), but there are also sizeable hippo populations in the Rufiji as it runs through Nyerere National Park, in the Nile below Murchison Falls, and in Rift Valley lakes such as Naivasha, Nakuru, Manyara and Baringo.

A hippo's 'yawn' is actually a threat display. (AVZ)

GIRAFFE

The world's tallest land mammal and heaviest ruminant, the giraffe (*Giraffa camelopardalis*) is instantly recognisable, with its elongated golden-maned neck and blotched coat. Standing up to 5.8m high and, exceptionally, weighing in at up to 2,000kg, the male is considerably larger than the female, but individuals of both sexes and all ages are very similar in shape. Most giraffes have dark brown spots or blotches all over their pale fawn bodies, except for on their bellies, with every individual having its own unique pattern. In some parts of East Africa, notably the Lake Manyara area, it is not uncommon to come across individuals so dark as to appear melanistic from a distance.

The giraffe favours savannah habitats scattered with medium-sized *Acacia*, *Commiphora* and *Terminalia* trees, feeding almost exclusively on leaves at heights of 2–6m, and using its versatile 40–50cm-long tongue to extend its reach even further. Its only competitor for this canopy-feeding niche is the elephant, but giraffes will sometimes also take leaves from lower down and will occasionally even graze on grass. A giraffe's heart, which weighs up to 10kg, has to generate a blood pressure double that of an average large mammal in order to pump blood to the brain via the long neck (which like all other mammalian necks, possesses just seven vertebrae).

A giraffe can run at speeds of up to 50km/h, revealing a rather odd 'slow-motion' gait when it does so. It also looks very clumsy when it drinks, having to spread its legs wide apart before lowering its neck to the surface. Although giraffes are often said to be mute, they do grunt and bleat occasionally, and – like elephants – most communication probably occurs at frequencies inaudible to the human ear. Male giraffes often engage in 'necking' behaviour, wherein two individuals spar by intertwining their necks, and striking each other with their heads in a show of

The Maasai giraffe is the commonest race and has the most irregular coat pattern. Youngsters must bend low to suckle. (AVZ)

Reticulated giraffes have the cleanest markings of all races. (AVZ)

strength. Necking has various functions, ranging from combat (occasionally to the death), to affection and sexual arousal – indeed, same-sex mounting in male giraffes is more frequent than heterosexual coupling.

Giraffes are modern representatives of what was probably the first ruminant family to colonise Africa's expanding grasslands some 20 million years ago. Numerous genera proliferated across the Old World tropics until about a million years ago, after which they fell into decline. Their demise might well have been exacerbated by human hunter-gatherers and reflects a general trend away from gigantism among other African ungulates. But the main factor behind giraffe extinction was undoubtedly the rapid rise and diversification of the more recently evolved bovids into most herbivore niches over the past few million years.

Formerly distributed throughout east and southern Africa, the giraffe is now more or less restricted to conservation areas, where it is generally common and easy to see in suitable habitat. It is absent from forest interiors and from a few specific savannah reserves, notably southern Nyerere, Ngorongoro Crater floor, and Queen Elizabeth National Park. Herds generally comprise about 15 members, which often disperse singly or in smaller groups, but aggregations of up to 50 animals are quite a common sight in some areas, notably Murchison Falls and northern Nyerere.

Scientists recognise up to 12 subspecies of giraffe, based largely on regional differences in coat pattern, of which three are present in East Africa. A recent genetics study has indicated several of these subspecies should be

WHAT'S IN A NAME?

The name giraffe almost certainly derives from Arabic, most likely *zirafah* ('the tallest') but also possibly *ziraafa* ('assemblage', in reference to its patchwork appearance) or *xirapha* ('swift walker'). In older English it was known as the 'camelopard' after the Latin *camelopardalis*, meaning leopard-like camel, and its Afrikaans name remains *kameelperd*! The giraffe's closest relative (indeed, the only other member of the family Giraffidae) is the okapi (*Okapia johnstoni*), a thick-necked, horse-sized, chocolate-coloured Congolese rainforest endemic, which eluded Western science until 1901, and whose range reputedly reached Uganda's Semuliki National Park until the 1930s.

treated as separate species, but this theory has yet to gain widespread acceptance. The most distinctive and handsome subspecies, sometimes treated as a full species, is the reticulated giraffe (*G. c. reticulata*) of northern Kenya, which has polygonal spots transected by a neat grid of bright white lines, and is quite common in Meru National Park and Samburu-Buffalo Springs. The most widespread East African subspecies is the Maasai giraffe (*G. c. tippelskirchi*), which has more irregular-looking blotches on a yellow-fawn background. This is replaced in Uganda and western Kenya by Rothschild's giraffe (*G. c. rothschildi*), which recent DNA studies suggest is genetically identical to the Nubian giraffe (*G. c. camelopardalisis*) of Sudan and Ethiopia. It can be seen in Uganda's Murchison Falls, Kidepo Valley and Lake Mburo national parks, as well as in Kenya's Lake Nakuru National Park.

Rothschild's giraffe is frequently seen in large groups on the grasslands of Murchison Falls National Park. (AVZ)

Buffalos frequently wallow in mud to stay cool and combat insect infestations, often with rather messy results. (AVZ)

BUFFALO, ANTELOPE AND ALLIES

Antelope are an unwavering feature of the African landscape, thriving in practically every habitat from the Congolese rainforest to the Sahelian desert, except where they have been eliminated by people. Indeed, despite the more charismatic likes of elephant, lion and giraffe, it is probably antelope that define the safari experience, with at least one species – usually more – being present everywhere. Perhaps this very abundance and ubiquity is why many safari guides treat the likes of gazelle and wildebeest as 'junk animals', but this view is very short-sighted: widespread herd grazers such as impala and kob display fascinating social behaviour and are very photogenic, while rarer or more localised species such as the magnificent greater kudu or elusive yellow-backed duiker often rank high on the wish list of the more experienced safari-goer.

Antelope are placed in the family Bovidae, which worldwide comprises almost 140 species of cloven-hoofed ruminant split across 50 genera, and includes such familiar farmyard animals as cattle, sheep and goats. Africa is the world's most important centre of bovid speciation, with more than half the world's species being endemic to the continent, and this diversity reaches its apex in East Africa, where approximately 45 species represent seven of the nine bovid subfamilies. The term antelope actually has a limited scientific taxonomic validity, since the spiral-horned antelope of the tribe Tragelaphini have a closer evolutionary relationship to cattle and buffalo (all are placed in the subfamily Bovinae) than to gazelles, duikers and other straight-horned antelope. Nevertheless, the term antelope is in such widespread colloquial usage that it would seem contrary to buck the convention here.

Africa's numerous antelope species are more closely related to sheep and cows than to the Eurasian deer, which are placed in the separate family Cervidae, a non-African lineage that differs from other ruminants in having bony seasonal antlers rather than permanent non-bony horns. Indeed, the similarities between antelope and deer represent a quite remarkable case of convergent evolution, one so thorough that it is seldom even recognised as such. The inability of the more recently evolved Cervids to cross the Sahara is a contributory factor to the immense bovid diversity in Africa as compared with, say, Asia, where deer and antelope are in direct competition.

Buffalo

The widespread African buffalo (*Syncerus caffer*) is Africa's only wild ox, and is closely related to the very similar-looking Indian water buffalo (and thus to domestic cattle). It is also the largest and most powerful African bovid, weighing up to 800kg, and is infamous for its unpredictable temperament – though attacks on people are actually very rare and are generally perpetrated only by wounded or surprised animals, with solitary bulls being particularly irascible. The imposing bulk of the adult African buffalo ensures it has few natural enemies, but it regularly falls prey to lions in some areas – especially the Maasai Mara, northern Serengeti and Ruaha, all of which are known for their large prides, though the big cats sometimes come off second best.

Buffaloes can adapt to most habitats, but are dependent on the availability of a reliable water source – primarily for drinking but also for bathing and wallowing.

A herd of African buffalo drinking at a waterhole. (AVZ)

Two subspecies are recognised, with the larger all-black savannah subspecies (*S. c. caffer*) being present in suitable protected areas throughout east Africa, though it is intermingled with genes from the chestnut-tinged forest buffalo (*S. c. nanus*) in parts of Uganda, notably Queen Elizabeth and Murchison Falls national parks. Very gregarious, the savannah buffalo is typically seen in mixed-sex herds of around 30–50 animals, but older males often roam around singly or in small bachelor groups after they have been ejected from a larger herd. With a bit of luck, you can expect good buffalo sightings on any East African safari, but certain parks are known for their large herds. These include Tsavo East and the Serengeti-Mara ecosystem, and especially Katavi National Park in western Tanzania, where several herds of 1,000–2,000 gather during the dry season.

The reddish tinge shows this buffalo to be a hybrid of the savannah and forest subspecies. (AVZ)

The eland is the world's largest antelope. (AVZ)

Spiral-horned antelope

Tragelaphini is a tribe of large to medium-sized antelope endemic to Africa, whose Asiatic ancestors arrived on the continent about 7 million years ago and most closely resembled the present-day bushbuck and nyala. The nine extant species, six of which are present in East Africa, are typically grey to chestnut-brown in colour, with a pattern of white spots and/or stripes on their flanks, face and forelegs, and, in several species, a pronounced dewlap and/or spinal crest. Males and females tend to look significantly different, with males generally being a head taller and sporting twisted or spiralled horns (present in both sexes with the eland). Like buffaloes, the tragelaphines are less fleet of foot than typical antelope, for which reason they are mostly associated with wooded habitats where stealth is a more important survival tool than flat-out speed. Oddly enough, however, the close genetic relationship between Tragelaphini and Bovini (cattle) is most apparent in the savannah-dwelling elands, which are relatively recently evolved cow-sized antelope whose appearance is pretty much what you'd expect were you to hybridise the two lines.

Eland

The common eland (*Tragelaphus oryx*) is Africa's largest antelope, with a shoulder height of 150–180cm and a highest recorded weight of 945kg. It is light tan-brown in colour, usually with a few faint white vertical stripes, and its somewhat bovine appearance is accentuated by the square contours and large dewlap. It could only conceivably be mistaken for the Derby's eland (*T. derbianus*), a west African species that is more brightly coloured and slightly smaller than the common eland (despite sometimes being referred to as the giant eland on account of its larger horns), and which no longer occurs in East Africa, having been hunted to extinction in north-western Uganda in the 1970s.

The eland is the most gregarious of the tragelaphines, typically seen in parties of around 10–15 animals, though herds of 50-plus are not uncommon and seasonal migrations may involve several hundred. It was revered by the hunter-gatherers who once inhabited much of eastern and southern Africa, and is the most commonly depicted animal on the rock paintings they left behind. Thinly but widely distributed in East Africa, eland are often very skittish and difficult to approach closely, and can jump up to 3m high from a dead start – both of which are important defence strategies for a relatively slow antelope of open habitats. Generally associated with grassland, montane moorland and lightly wooded savannah, the eland might be seen in almost any game reserve across the region, but Serengeti and Samburu are among the more reliable spots.

Kudus

Arguably the most magnificent of African antelope, the greater kudu (*Tragelaphus strepsiceros*) is second in stature only to the eland, standing up to 155cm high at the shoulder and weighing up to 320kg. Generally associated with thick woodland and often seen along watercourses, it has a grey-brown coat and up to ten vertical white stripes on each flank, but is most notable for the statuesque male's double-spiralled horns, which can grow to be 1.8m long. This shy species generally travels

A male greater kudu has impressive corkscrew horns. (AVZ)

in mixed- or single-sex herds of up to ten animals, though males are often seen alone or in pairs. Although it is not an especially fast runner, it is an accomplished jumper, capable of clearing fences twice its shoulder height. Greater kudu are common in suitable habitats in southern Africa, but the East African population has never recovered from the hammering it took during the rinderpest epidemic in the late 19th century, and it is only likely to be seen in Kenya's Lake Bogoria National Reserve, and in southern Tanzanian national parks such as Katavi, Ruaha, Mikumi and Nyerere.

The lesser kudu is smaller and daintier than the greater. (AVZ)

Endemic to East Africa and southern Ethiopia, the lesser kudu (*T. imberbis*) is a thinly distributed and skittish antelope that often occurs alongside the greater kudu in dry acacia woodland. It can be distinguished from the greater kudu by its smaller size (shoulder height 100cm), two white throat patches and having more vertical stripes (at least 11). Nowhere common, you might come across small, shy herds in Tsavo, Samburu or Ruaha national parks.

Bushbuck

Endearingly Bambi-like, the bushbuck (*T. scriptus*) is possibly the most widespread medium-sized antelope in East Africa, present in all non-arid habitats, and still quite common outside protected areas. Its attractive coat shows great regional variation in colouring, with the male being dark brown or chestnut (in parts of

As in all tragelaphine antelope, only the male bushbuck has horns. (AVZ)

Male sitatunga (AVZ)

Ethiopia black) while the much smaller female is generally pale red-brown. The male has relatively small, straight horns for a *Tragelaphus* antelope. Both sexes have white throat patches and are marked with white spots and sometimes stripes. Most common in forest and riverine woodland, where it occurs singly or in pairs, the bushbuck is harder to see than you might expect of such a common antelope, tending to be secretive and skittish except where it is used to people.

Sitatunga

The semi-aquatic sitatunga (*T. spekii*) is a widespread but seldom seen inhabitant of swampy habitats from Botswana to Sudan. It looks rather like a bushbuck but is significantly larger and shaggier, with a spread-eagled stance and unique splayed hooves that assist it in manoeuvring through its watery home. The best place to see one is Tanzania's Rubondo Island National Park, but it also quite easily observed in Saiwa Swamp National Park (Kenya) and the Katanga Wildlife Reserve (Uganda).

Bongo

The bongo (*T. euryceros*) is a massive (up to 400kg) antelope, whose deep russet coat is marked with around a dozen bold white vertical stripes, and whose

main centre of distribution is the lowland rainforests of west-central Africa. The East African subspecies (*T. e. isaaci*), effectively a Kenyan endemic since the last Ugandan specimen was shot circa 1914, is restricted to three isolated montane forest pockets, of which the Aberdares supports a barely viable population of around 50 individuals; the situation in the unprotected forests of the Mau Escarpment is uncertain; and no definite records exist for Mount Kenya since the mid-1990s – though attempts to reintroduce zoo-bred specimens of this endangered race are under way. The first East African record of lowland bongo

The elusive bongo is a very rare sight in East Africa. (AVZ)

(*T. e. euryceros*), a widespread west-central African subspecies, was camera-trapped in Uganda's Semuliki National Park in 2018.

Sable and roan antelope

The rather horse-like antelope of the genus *Hippotragus* consist of two living species, the sable antelope (*H. niger*) and the roan antelope (*H. equinus*), and once also included the bluebuck (*H. leucophaeus*), which was hunted to extinction within 150 years of European settlement in South Africa. Uncommon in East Africa, the roan antelope stands up to 1.5m tall at the shoulder, has a uniform fawn-grey coat with a black-and-white-striped face, tasselled ears, short decurved horns and a light mane. It is usually seen in shy herds of 5–10 animals in dry woodland, and is thinly distributed in southern Tanzania, very rare in the Serengeti, rather more common in Kenya's Ruma National Park, and extinct in Uganda.

Male sable antelope (AVZ)

The more striking male sable is among the most handsome of African antelope, its jet-black coat off-set by a white face, underbelly and rump, and decurved horns up to 1.3m long. The female, which is chestnut brown and has shorter horns, could be mistaken for a roan, but has a more defined white belly and lacks the roan's facial markings The 8,000 sable antelope that migrate through the greater Nyerere-Niassa ecosystem in southern Tanzania constitute by far the largest wild population anywhere in Africa, but are mainly concentrated south of the river and are thus seldom seen by tourists. Elsewhere, the sable is an uncommon resident in Ruaha, Katavi and Saadani national parks (Tanzania), but the only place where a sighting is guaranteed is Kenya's Shimba Hills National Reserve, which protects a herd of about 120 habituated animals.

The roan antelope is larger and paler than the sable, with shorter horns and distinctive tassels on its ears. (AVZ)

The lethal horns of the beisa oryx are an effective defence against even lions (AVZ)

East African oryx

The East African oryx (*Oryx beisa*) is an unmistakable dry-country antelope that stands 120cm high at the shoulder and has striking straight horns that can exceptionally reach 1.2m in length, sweeping backwards at the same angle as the forehead and muzzle. Powerfully built, with a deep chest and thick neck, it has an ash-grey coat, bold black facial marks, and generally travels in small nomadic herds of up to ten animals. An oryx drinks readily where water is available, but it can go without for almost as long as a camel, obtaining all its needs from the plants it eats, and – in order to survive the extreme heat of its semi-desert habitat – allowing its body temperature to rise as high as 45°C without perspiring.

Two subspecies are recognised: beisa oryx (*O. b. beisa*) and fringe-eared oryx (*O. b. callotis*). They are now regarded to be conspecific with the gemsbok (*O. gazella*) of southern Africa. The beisa oryx is common in arid country north of the Tana River, especially in Samburu-Buffalo Springs, while the fringe-eared oryx is most likely to be seen in Tsavo East, though it also occurs in small numbers in northern Tanzania (Tarangire, Mkomazi and the northern Serengeti).

Wildebeest, impala and allies

Alcelaphinae is a subfamily of mostly large and rather ungainly-looking grazers, characterised by their short horns, sloping backs and elongated faces. The best known is the blue wildebeest (or brindled gnu), but it also includes the various hartebeests and the atypically handsome impala.

Blue wildebeest

Among the best known of African antelope, the blue wildebeest or gnu (*Connochaetes taurinus*) stands up to 1.5m tall and has a dark grey-brown coat, which can make it look rather buffalo-like from a distance (indeed, the name *wilde beest* is Afrikaans/Dutch for 'wild ox'). At closer quarters, however, its slighter build, paler coloration

Blue wildebeest run the gauntlet of crocodiles as they cross the Mara River during their annual migration. (AVZ)

Plains zebra often associate with blue wildebeest herds. (AVZ)

and shaggy beard should preclude confusion. This species is common to abundant in most grassland habitats south of the equator, and herds of 20–100 are a regular sight in the northern Nyerere, Ruaha and most other Tanzanian reserves, often in the company of plains zebra.

Mysteriously, however, the wildebeest is totally absent from the northern hemisphere (ie: most of Kenya and all of Uganda), this despite the presence of almost 2 million in the Equator-nudging Serengeti-Mara ecosystem, the site of Africa's greatest extant large mammal migration. Here, every year, more than 2 million ungulates – predominantly blue wildebeest, but also zebra, gazelle and eland – follow the rains in a reasonably (but far from infallibly) predictable cycle. They calve in the south-eastern Serengeti over December to May, then march in a column of up to 40km long to cross the Grumeti River in June/July, before dispersing into the northern Serengeti and Maasai Mara from July to October, from where they plod back to the southern Serengeti to arrive in late November, only for the cycle to start all over again.

Wildebeest have a distinctive long nose. (AVZ)

STAKING A CLAIM

Inoffensive and placid as they might appear, antelope are among the most territorial of creatures, with males in particular adopting a number of strategies to mark their ground and defend it against rivals. The most obvious clue to this aspect of antelope behaviour is sexual dimorphism in horn size, which suggests that these sabre-like appendages play a greater role in male territorial fights and dominance displays than in defence against predators. Other aspects of territorial behaviour, such as the staring down of rivals, often while standing in an exaggeratedly erect posture, are also common to most antelope species.

Other traits are peculiar to certain types. Dik-dik, oribi and klipspringer, for instance, have pre-orbital glands (in front of their eyes) whose secretions are used to scent-mark their territory, while male impalas defecate and urinate on strategically placed middens to the same end. A male topi will often stand prominently on a termite mound to advertise his dominance, whereas sable and roan antelope advertise their status by thrashing shrubs and trees around with their long horns. The exact behaviour might differ from one species to the next, but any observer who is patient enough to spend time watching an antelope herd is bound to observe this sort of territorial behaviour and interaction.

Hartebeest and topi

Often seen standing atop a termite mound scanning the surrounding plains for predators or displaying itself to rivals, the hartebeest (*Alcelaphus buselaphus*) is similar in height to the blue wildebeest, and its narrow elongated face gives it if anything an even more morose demeanour. It has a more slender build, however, and is readily identified by the combination of large shoulders, pale yellow-brown coat and smallish horns in both sexes. The half a dozen or so different subspecies across Africa are best distinguished from each other by geographical distribution and horn shape, but all are generally seen in small family groups in grassland or light woodland. Most common in East Africa is the kongoni or Coke's hartebeest (*A. b cokei*) of the Serengeti-Mara ecosystem. It is replaced by the uncommon Lichtenstein's hartebeest *A. (b.) lichtensteinii* in southern Tanzania (considered by some authorities to be a separate species) and by Jackson's hartebeest

Jackson's hartebeest (AVZ)

A. b. jacksoni in northern Kenya and Uganda, which is easily observed in the north of Murchison Falls National Park.

The topi (*Damaliscus lunatus*) is a darker, glossier and more handsome variation on the hartebeest, and tends to be seen in similarly open habitats in similarly sized herds. It is dark brown in general coloration, with a purplish-black sheen on the flanks and snout, and striking yellow lower legs. Widespread but patchily distributed, the topi is most common in the Serengeti-Mara ecosystem, where it occurs alongside Coke's hartebeest.

The hirola (*Beatragus hunteri*), also known as Hunter's hartebeest, resembles a downscaled hartebeest with impala-like horns and distinctive white 'spectacles'. Listed by the IUCN as Critically Endangered, the hirola is naturally confined to a restricted area of grassland in southern Somalia, where it is probably extinct, and the northern coastal belt of Kenya, where the population plummeted from 10,000-plus in the mid-1970s to perhaps 300 today, largely as a result of drought and competition with livestock. The numeric decline has probably been stemmed – at least temporarily – by the creation of the 530km^2 Arawale Hirola Reserve in 1974 in the northeast of Kenya, though this nominally protected population remains highly vulnerable. No significant hirola population is known to exist in captivity, but two small herds introduced to Tsavo East National Park in 1963 and 1996 have adapted well to this semi-arid habitat a short distance south of their natural range.

Topis thrive in open grassland. (AVZ)

Only male impala have horns. (AVZ)

Impala

The impala (*Aepyceros melampus*) is a slender, medium-sized (shoulder height 90cm) antelope that bears a strong superficial resemblance to the gazelles, but is in fact more closely related to hartebeest and wildebeest – indeed, it is often lumped in with Alcelaphinae, though most authorities now place it in the monospecific subfamily Aepycerotinae. Larger and more elegantly proportioned than any gazelle, the impala is a graded fawn to chestnut in colour, with diagnostic black and white stripes running down its rump and tail. The female is hornless, but the male's black, annulated, lyre-shaped horns can be magnificent, particularly in the north of its range.

Highly gregarious and equally territorial, the impala is generally seen in hundred-strong herds of females and young presided over by a single dominant male, or in smaller bachelor herds of 10–30 animals. It is a wonderfully agile leaper, capable of high jumps of 3m and long jumps of more than 10m, and a disturbed herd will often spring in all directions as a strategy to confuse predators. Often, every impala in a herd will perform an exaggerated leap as it files across a road or a clearing. This showy behaviour is thought to have developed as a way of demonstrating fitness to any watching predator.

The impala is the most numerous antelope in many sub-equatorial African savannah reserves, particularly in southern Tanzania, where it inhabits lightly wooded and woodland-edge habitats. It is also reasonably common in bushy parts of the Serengeti-Mara ecosystem, while the only extant Uganda population persists in Lake Mburo National Park. Impalas are scarce north of the equator, with just a handful of relict populations present in northern Kenya.

Impala are both grazers and browsers, so prefer a good mix of grass and trees. (AVZ)

Defassa waterbuck have full white rumps. (AVZ)

Kob, reedbuck and allies

The subfamily Reduncinae comprises eight species of medium-large antelope, most of which are associated with moist grassland habitats.

Waterbuck

The largest and most distinctive of the reduncines is the waterbuck (*Kobus ellipsiprymnus*), which has a shoulder height of up to 135cm and is easily recognised by its shaggy grey-brown coat and the male's large lyre-shaped horns. The Defassa waterbuck (*K. e. defassa*), of the Rift Valley and areas further west, is a more chestnut subspecies with a full white rump, while the nominate eastern subspecies has a white ring around its rump, as though it had just sat on a freshly painted toilet seat. Waterbuck are almost always seen in the vicinity of standing water, where herds of up to ten individuals are presided over by one (often very aggressive) male. Both races are widespread in suitable habitats throughout East Africa, but are especially common in Queen Elizabeth, Murchison Falls, Akagera, Arusha, Lake Nakuru, Crescent Island and Lake Manyara national parks.

Common waterbuck is distinguishable by the white ring on its rump. (AVZ)

Top and above: One of the most handsome East African antelopes is the Uganda kob, which forms large breeding harems (leks) in grassy floodplains in Uganda. Subdominant males stand sentry on the periphery. (AVZ) *Right:* Within East Africa, the puku is restricted to southern Tanzania. (AVZ)

Kob and puku

Uganda's national antelope is the near-endemic Uganda kob (*Kobus kob thomasii*), a subspecies of the West African kob whose range is confined to grassy floodplains and open vegetation near water in Uganda and the southern Sudan. This species has a reddish-brown, impala-like coloration, but is bulkier in appearance and lacks the impala's black rump stripes. Highly gregarious and territorial, the Uganda kob is most abundant in Queen Elizabeth and Murchison Falls, where herds of several hundred individuals congregate in cropped grassy breeding grounds known as leks. The kob has a complex social structure: the dominant male in any herd is supported by several younger males, which spend most of their time keeping guard on the verge of the main group, and emit a loud snorting bark when they sense danger – behaviour that often inadvertently assists safari vehicles in locating lions. The southern equivalent of the kob, the similar-looking puku (*Kobus vardonii*) is a rather localised floodplain resident. Its range is centred on the Kilombero Valley in the Nyerere-Niassa ecosystem, which supports around 40,000 animals, representing about half of the global population.

A bohor reedbuck is well camouflaged in dry grassland. (AVZ)

Reedbuck

Arguably the most nondescript of all Africa's medium-sized antelope, the three reedbuck species of the genus *Redunca* are all lightly built, pale grey or tan grassland dwellers, with white underbellies and short horns. They are most often seen in pairs or trios that tend to flee at the slightest disturbance. The Bohor reedbuck (*R. redunca*), which has very short forward-curving forms, is the only species likely to be seen in East Africa's non-montane habitats, except for in southern Tanzania, where its range overlaps with that of the larger and more straight-horned southern reedbuck (*R. arundinum*). Chunkier and greyer in appearance than its lowland counterparts, the mountain reedbuck (*R. fulvorufula*) is widespread in East Africa, but is almost exclusively associated with grassy and rocky mountain slopes.

Duikers

The subfamily Caphalophinae comprises 18 species of smallish antelope, all but one of which inhabit forest, where they tend to vanish quickly into the undergrowth, hence the common name duiker, which is Afrikaans for 'diver'. The one exception to this rule is the common or grey duiker (*Sylvicapra grimmia*), a common resident of savannah habitats that might be seen practically anywhere in East Africa excluding forest interiors and deserts. Standing about 50cm high and almost invariably seen singly or in pairs, the common duiker holds itself more like a steenbok than any other duiker, but it is greyer in colour, and can be distinguished from other small savannah antelope by the black tuft of hair that sticks up between its horns.

The hunchbacked forest duikers of the genus *Cephalophorus* are probably the least understood of African antelope, and few tourists will encounter any of the region's eight species. In Kenya and Tanzania you might come across the 45cm-tall Harvey's red duiker (*C. harveyi*), which is deep chestnut in colour with a white tail and black patch on its snout, and occasionally appears in coastal forest clearings in the likes of Saadani and Shimba Hills. It is replaced by the very similar Natal red duiker (*C. natalensis*) on the south coast of Tanzania; by Peters's red duiker (*C. callipygus*) in western Kenya and Uganda; and by the endemic Rwenzori red duiker (*C. rubidus*) on the lower Rwenzori. Also placed in the red duiker cluster, the endangered Aders's duiker (*Leucocephalophorus adersi*) is near-endemic to Zanzibar, where no more than 1,400 individuals survive, mostly in Jozani Forest.

Abbott's duiker (*Cephalophus spadix*) is a relatively large species (shoulder height 75cm), which, having recently become extinct in Kenya, is now confined to five forested montane 'islands' in Tanzania, namely Kilimanjaro, Usambara, Udzungwa, Uluguru and Rungwe. Along with Aders's duiker, it is the most threatened of African duikers, with a total population estimated at 1,500, but is quite often glimpsed by hikers on Kilimanjaro and can be identified by its glossy off-black torso, red forehead tuft and large size. The western equivalent, recorded in several forests in western Uganda and Rwanda, is the yellow-backed duiker (*C. silvicultor*), which is heavier than a bushbuck and is sometimes encountered fleetingly along the forest track leading uphill from Buhoma in Bwindi Impenetrable National Park. At the other end of the size scale, the widespread but shy blue duiker (*Philantomba monticola*) is the tiniest East African antelope, with a typical shoulder height of 35cm and weight of 5kg. This species, which sometimes falls prey to the crowned eagle, is most common in coastal thickets and forest, but occurs throughout the region in suitable habitats.

Left: Common duiker (AVZ) *Right:* The Natal red duiker, *Cephalophus natalensis*, is associated with dense forested habitats along the east coast of Tanzania. (AVZ)

Grant's gazelle has longer horns than the 'Tommy', and frequently uses them in combat. (AVZ)

Gazelles

Antilopinae is the most diverse antelope subfamily, represented by 14 genera, several of which are distributed outside of Africa. Best known are the gazelles of the genera *Eudorcas*, *Nanger* and *Gazella*, a group of graceful, relatively small herd antelope associated with arid and grassland habitats, and represented by around 20 species worldwide.

'Tommies' and Grant's

The nippy Thomson's gazelle (*E. thomsonii*), popularly known as 'Tommy', and Grant's gazelle (*N. granti*) are among the most characteristic grazers of the Serengeti-Mara ecosystem, where they are particularly favoured as prey by the even speedier cheetah. Though superficially similar, with their tan brown upperparts and white belly, the two gazelles are quite easy to tell apart: Grant's gazelle is much larger and has bigger horns than the Tommy, but lacks the latter's bold black horizontal flank stripe. Thomson's gazelle is endemic to East Africa, with a range centred on the Serengeti-Mara ecosystem, extending into the nearby Amboseli and Arusha national parks. Grant's gazelle occurs alongside the 'Tommy' in all of these places, but its range also extends southward through central Tanzania as far as Ruaha National Park. In 2021, two former subspecies of Grant's gazelle were accorded full species status. These are Bright's gazelle (*Nanger notatus*), which can be seen in Samburu-Buffalo Springs and elsewhere in northern Kenya, and Peters's gazelle (*N. petersii*), whose range is focused on the far east of Kenya.

Two male Thompson's gazelles locked in territorial combat. (AVZ)

The gerenuk uses its long neck and hind legs to reach the leaves that most rival browsers cannot. (AVZ)

Gerenuk

One of East Africa's most striking antelope, the gerenuk (*Litocranius walleri*) looks like a cross between a gazelle and an impala, but can be distinguished from either by its extraordinarily long neck (as reflected in the Swahili name *swala twiga*, meaning 'gazelle giraffe') and unique habit of feeding from acacia trees standing on its hind legs, neck at full stretch. Associated with dry acacia savannah, it occurs in a few Tanzanian reserves, including Mkomazi and Tarangire, but is easier to see in Kenya, particularly in Tsavo East and Samburu-Buffalo Springs.

Unlike their larger cousins, small antelope tend to be shy, jumpy and unsociable, and are usually seen singly or in pairs. An exception is the oribi (top, AVZ), which is the largest of the so-called dwarf antelope and often seen in small family groups – especially in the open grasslands of Uganda's Murchison Falls national park. The steenbok (left, AVZ) is the most nondescript of the dwarf antelope and resembles a miniature oribi. Kirk's dik-dik (bottom left, AVZ) is easily recognised by its bold pale eye and twitchy nose. The klipspringer (bottom right, AVZ) is the only species that habitually clambers around rocks, where its rubbery hooves give it remarkable sure-footedness.

Dwarf antelope

The other East African representatives of the subfamily Antilopinae are placed in the tribe Neotragini, and are collectively referred to as the dwarf antelope, in reference to their diminutive size.

The largest member of the Neotragini is the oribi (*Ourebia ourebi*), a widespread but often uncommon grassland antelope that stands up to 65cm high at the shoulder and is sandy in colour, with a white belly, small straight horns in both sexes, and a distinctive round, black glandular patch below each ear. It is typically seen in pairs or small herds in open grassland habitats, most commonly in the northern part of Murchison Falls National Park and in Rwanda's Akagera National Park, where it tends to draw attention to itself with a trademark sneezing alarm call.

Common in northern Tanzania and parts of Kenya, the smaller steenbok (*Raphicerus campestris*) has fawn upper parts, clear white underparts, large ears and short straight horns. It is often encountered in pairs and tends to 'freeze' when disturbed. Bates's pygmy antelope (*Nesotragus batesi*) is the second-smallest African ungulate, a Congolese rainforest species that has been recorded in forest interiors in Uganda's Semuliki and Queen Elizabeth national parks.

The klipspringer (*Oreotragus oreotragus*) is an intriguing species associated almost exclusively with rocks (its name is Afrikaans for 'rock jumper'). It boasts several unusual adaptations to its favoured habitat of rocky slopes or koppies, notably the unique ability to walk on its rubbery hoof tips, and its coarse but hollow fur providing good insulation at high altitude. It also possesses binocular vision, a feature more normally associated with carnivores than herbivores, presumably to help it gauge jumping distances accurately. Almost always seen in pairs, which bond for life, this perky little antelope has a uniform grizzled grey or grey-brown coat, short straight horns, and an arched back reminiscent of a duiker. Conservative anatomic features suggest it represents one of the most ancient of antelope lineages, having evolved from the earliest African antelope stock some 10–15 million years ago, possibly in the Ethiopian highlands. Today, it occurs in suitable habitats throughout East Africa, and it is often seen in the Lobo region of the Serengeti and on rocky slopes associated with the Rift Valley.

Among the most distinctive and common dwarf antelope are the dik-diks of the genus *Madoqua*, two species of which are present in East Africa. Kirk's dik-dik (*M. kirkii*) has the more southerly distribution, being a common resident of most reserves south of Mount Kenya, notably Arusha National Park, while the more northerly Guenther's dik-dik (*M. guentheri*) is associated with the arid badlands of northern Kenya, with the ranges of the two overlapping (and possibly a degree of hybridisation occurring) in Samburu-Buffalo Springs. The two species are practically impossible to tell apart by sight, but are easily distinguished from other small antelope by their grizzled grey-brown coat, prominent white rings around the eyes, and twitchy elongated nose which is thought to play a role in regulating temperature. Dik-diks are almost always seen in pairs. They can be skittish, but are also often very relaxed photographic subjects, provided one approaches them slowly and waits a few minutes for them to settle down.

The rock hyrax, as its name suggests, occurs on rocks, often around cliffs and gorges. (AVZ)

SMALL MAMMALS
HYRAXES

Confined to Africa, except for a small population in Arabia, hyraxes are superficially rodent-like mammals that somewhat resemble an overgrown guinea pig but are larger and sharper of tooth. A bizarre truism, much beloved of safari guides, is that the hyraxes' closest living relatives are elephants. This factoid loses some of its 'golly gosh!' value when you realise that the elephant/hyrax split dates back almost as far as the split between cats and people. In fact, modern hyraxes are dwarfish relicts of a near-ungulate order that dominated the African herbivore niche about 35 million years ago, when some species were as large as horses. They still possess several 'early mammal' characteristics such as a very long gestation period (up to eight months) measured against size, and poorly developed internal temperature regulators, which requires them to spend a lot of time basking in the sun or huddled in groups.

The most common species today are the practically indistinguishable rock hyrax (*Procavia capensis*) and bush hyrax (*Heterohyrax brucei*), which typically lives in rocky or mountainous habitats, forming strongly territorial family groups of around 10–20 individuals that can become very tame when they are used to people, for instance at Seronera and Lobo lodges in the Serengeti. The less common and seldom observed tree hyrax (*Dendrohyrax arboreus*) is a nocturnal forest creature that announces its presence with an unforgettable banshee wail. This terrifying sound is often heard in the Budongo Forest in western Uganda.

AARDVARK

The aardvark (*Orycteropus afer*), weighing 40–70kg, is really too large to warrant the label 'small mammal'. Yet finding a suitable category for this bizarre creature is not easy. As the only living species in the order Tubulidentata, it has highly uncertain evolutionary affinities and is surely the oddest of all African mammals. Its elongated snout and insectivorous habits bring to mind the South American anteaters, yet as its Dutch-derived name (literally 'earth pig') indicates, it also bears some similarity to domestic pigs in its shape, size and naked pinkish skin. Other features include a heavy kangaroo-like tail, long upright ears, and muscular long-nailed feet with which it burrows into termite mounds, using its long sticky tongue to snaffle as many as 50,000 termites in one night. Unusually for a mammal, its teeth lack a pulp cavity and enamel coating, and grow continuously.

One of the most eagerly sought species by repeat safari-goers, the aardvark is a widespread but uncommon inhabitant of savannah and arid habitats in sub-Saharan Africa. Being a shy animal that emerges to forage only late in the night, sightings are very rare. Solitary by nature, it feeds mainly on colonial ants and termites, and is most common in areas that are well endowed with termite mounds. It is not confined to protected areas, however, and does well in pastoral land where overgrazing by cattle and goats creates good feeding conditions.

The unique and elusive aardvark is most likely to be seen after dark on game drives through dry grassland savannah habitats. (AVZ)

The armoured scales of the ground pangolin are impenetrable to most predators. (AVZ)

PANGOLINS

Not so much similar to the aardvark as equally dissimilar to anything else, the pangolins or scaly anteaters of the order Pholidota are unobtrusive and seldom observed nocturnal insectivores whose distinctive scaled armour plating could cause them to be confused with a deviant form of reptile. The name pangolin derives from the Malay *penguling* meaning 'thing that rolls up', a reference to their tendency to roll up into an artichoke-like ball when disturbed. As with aardvarks, the taxonomic affinities of pangolins are uncertain, but a superficial resemblance to the American armadillos is probably a case of convergent evolution, and recent DNA tests suggest a closer relationship to the carnivores.

Eight pangolin species are recognised, divided equally between Asia and Africa, all of which are weak of tooth but powerful of claw – an ideal combination for an exclusive diet of ants and subterranean termite colonies. The ground pangolin (*Smutsia temminckii*) is most widespread in East Africa, though the much larger (35kg) giant pangolin (*S. gigantea*) also occurs in the forests of Uganda. Sightings are an extremely rare event, more because pangolins are highly secretive and nocturnal than because they are especially scarce, though the combination of deforestation and human predation has caused the giant pangolin to be listed as Vulnerable by the IUCN.

RODENTS

An estimated 40% of all mammal species in East Africa, and globally for that matter, belong to the order Rodentia, a diverse group of small to medium-small herbivores and omnivores distinguished by their prominent gnawing teeth. There are estimated to be some 2,500 rodent species worldwide, and they (along with bats) are the only orders of placental mammals that established themselves naturally (ie: without human influence) on Australia, where they comprise 25% of the indigenous mammalian species. An association between house rats and disease means that rodents are often thought of as dirty and rather undesirable, an unfair prejudice that makes little sense in a non-urban setting such as the African bush. All the same, the majority of rodents are small, nondescript and of little interest to the average safari-goer. The following account confines itself to a few of the larger and more distinctive varieties.

Porcupines

Porcupines (family Hystricidae) are represented in East Africa by two practically indistinguishable species. They are among the largest of rodents (up to 27kg) and have a unique covering of long black-and-white quills that protect them from predators and also occasionally betray their presence by rattling as they walk. Nothing more than modified hairs, these quills cannot be fired at will, nor do they contain any poison, but their sharp ends nevertheless form a powerful deterrent to any predator foolish enough to try it on. The quills do detach easily, which is why they are far more likely to be seen scattered around in the bush than the animal itself, which is widespread in East Africa but almost entirely nocturnal and very seldom observed. Porcupines will sometimes enter suburban gardens to root out bulbs and can damage orchards by ringbarking trees.

Despite its fearsome quills, the Cape porcupine (*Hystrix africaeaustralis*) is an extremely shy nocturnal creature that tends to flee from people. (AVZ)

Springhare

Another large and seldom seen nocturnal oddity is the East African springhare (*Pedetes surdaster*), which, despite its misleading name, is a true rodent. It is generally associated with dry savannah habitats, and is generally more common in southwest Africa than East Africa, though small numbers are present in the Serengeti and southern Tanzania. Larger and redder than any true hare, the most striking feature of the springhare is its kangaroo-like mode of locomotion – indeed, there are few odder (or more distinctive) sights on a night drive than a springhare's eyes bouncing up and down in the spotlight.

Squirrels

East Africa also supports a wide diversity of squirrels, most of which are associated with woodland or forested habitats, most commonly bush squirrels (*Paraxerus* spp), which are rusty brown in colour with a silvery black back and white eye-rings. Also likely to be seen on safari is the endearing unstriped ground squirrel (*Xerus rutilus*), a terrestrial dry-country creature with a grey-brown coat, a prominent white eye-ring, a silvery black tail, and a habit of standing on its hind legs holding food in its forepaws.

Unstriped ground squirrel in Tarangire National Park, Tanzania. (AVZ)

RABBITS

Rabbits and hares, though superficially rodent-like, belong to the Lagomorpha, a globally widespread order that has been around for at least 60 million years, but whose affinities remain a puzzle to taxonomists. Of the three East African species, the most widespread are the similar-looking scrub hare (*Lepus saxatilis*) and Cape hare (*L. capensis*), both of which are frequently seen in grassland and lightly wooded habitats towards dusk, and tend to freeze for a few seconds when disturbed, bounding away suddenly if they think they have been noticed. Restricted to the western part of the region, the smaller and shorter eared Bunyoro rabbit (*Poelagus marjorita*) is quite common in grassland habitats in Uganda.

Scrub hare (AVZ)

INSECTIVORES AND SENGIS

The order Eulipotyphla is low-key even by comparison with the rodents, this despite the presence of almost 200 described species continent-wide, of which three-quarters are mouse-like shrews of the family Soricidae. The most distinctive insectivores in East Africa are the hedgehogs of the genus *Atelerix*, inconspicuous and shy nocturnal creatures that for some reason seem to be observed quite regularly at Kirurumu Tented Camp and vicinity on the Rift Valley escarpment overlooking Lake Manyara. Endemic to Africa, and not members of Eulipotyphla, despite their insectivorous diet, the 15 species of sengi or elephant shrew belong in the ancient order Macroscelidea. These peculiar shrew-like creatures move like miniature kangaroos and have absurdly elongated twitchy noses. Secretive and largely nocturnal, they are occasionally seen around camps at dusk or on night drives, with the smaller species generally being associated with savannah habitats. The 'giant' sengis of the genus *Rhynchocyon* are colourfully patterned forest inhabitants that weigh 500g and are quite often glimpsed on forest paths, but usually make a quick getaway. Five species are now recognised. Most common and widespread are the recently split chequered sengi (*R. cirnei*), which occurs in southern Tanzania, and Stuhlmann's sengi (*R. stuhlmanni*), a Congolese species whose range extends into Uganda. More localised are the Zanj sengi (*R. petersi*) of the Tanzania coastal belt and Zanzibar, the golden-rumped sengi (*R. chrysopygus*) of the Kenya coast (sometimes seen at the Gedi Ruins near Watamu), and the recently (2008) discovered grey-faced sengi (*R. udzungwensis*) of the Udzungwa Mountains.

The large ears and 'nose-leaf' of the yellow-winged bat help it to receive echolocation signals, which reveal the whereabouts of flying insect prey. (AVZ)

Bats

Bats are perceived as harbingers of good fortune in many Asian and Middle Eastern cultures, but many Westerners fear them to the point of phobia, especially when they swish around hotel balconies at dusk. In reality, however, bats are an immensely important feature of any natural ecology, especially when it comes to the control of flying insect populations. Their legendary status as bloodsuckers is also irrelevant in an Old World context, the world's only three species of 'vampire' bat being confined to the Americas.

Placed in the order Chiroptera and traditionally aligned with the insectivores, bats are the only mammals that possess true wings, and the only ones capable of powered flight (as opposed to gliding). These wings are modified forelimbs that resemble outstretched human hands, with the digits connected by a fine membrane of webbed skin (the name Chiroptera derives from the Greek *cheir pteron* meaning 'hand wing'). In terms of diversity, Chiroptera is the second most successful mammalian order after rodents, with more than 1,000 species recognised globally, of which 20% occur on mainland Africa. It also has the widest distribution of any mammal order thanks to its aerial mobility.

Chiroptera was formerly divided into two suborders – Megachiroptera (fruit bat) and Microchiroptera (insect-bats). The names were somewhat misleading, since several 'mega' species are smaller than the largest 'micro' species. The two groups were more meaningfully distinguished from each other on the basis of morphology, habits and diet. As a rule, 'micro' bats are insectivorous, they have small eyes, complex ears, furless bellies and clawless forelimbs, and rely primarily on echolocation for navigational and hunting purposes. By contrast, almost all 'mega' bats are fruit-eaters, they have large eyes, simple tunnel-shaped ears, furry bellies and a claw on the second toe of the forelimb, and they rely on sight and smell to navigate and find food.

So marked is the apparent distinction between insect and fruit bats that it has long been suggested they represent completely different (albeit strongly convergent) evolutionary lines and should be placed in completely separate orders. However, genetic evidence has now shown that this is not the case. Chiroptera is now generally divided into two suborders – Yangochiroptera, which includes most insect-eating bats, and Yinpterochiroptera, which contains the fruit bats but also five families of insect-eating bats.

Insect-bats

The most interesting feature of insect-eating bats is their ability to fly at high speeds in dark caves or at night, and find and catch prey using only echolocation, a technique that cetaceans (dolphins and whales) use for catching fish and squid. This works by the bat emitting a mixture of modulated and constant sound signals, which bounce back to it from inanimate or other moving objects, allowing the flying bat to build up a continual sonic 'picture' of its environment and adjust its flight accordingly. This ability to instantly 'see' one's surroundings purely through sonic echoes is almost incomprehensible to the sight-oriented human mind, but doubtless an avian raptor's ability to capture flying prey using purely visual impulses would seem equally incredible to a creature that operated purely by echolocation.

More than 80% of described bat species are insect-bats. They are more widespread than fruit bats, since certain species have the ability to hibernate, allowing them to tolerate chillier climates – indeed, some five species are indigenous to Alaska! Close on 100 insect-bat species occur in equatorial Africa, including the common slit-faced bats (*Nycteris* spp.), leaf-nosed bats (*Hipposideros* spp.) and house bats (*Scotophilus* spp.), and you're likely to see a few fluttering around the bush at dusk wherever you go, although distinguishing the different species in the field requires some expertise (and, often, a specimen in the hand). One of the more conspicuous species is the yellow-winged bat (*Lavia fons*), a widespread resident of savannah and woodland whose large golden ears and yellow wings can make it look rather like a misplaced autumnal leaf when seen hanging in trees by day.

Fruit bats roost colonially in tall trees by day, but forage over long distances at night. (AVZ)

Fruit bats

More or less confined to the Old World tropics, Australasia and oceanic islands in between, fruit bats are well represented in East Africa, where they are associated with forested habitats and most species roost colonially in trees. The largest species to occur in the region is the Seychelles flying fox (*Pteropus seychellensis*), an oceanic island species that reaches the most easterly extent of its range on the Mafia archipelago, has a wingspan of around 1m, and weighs up to 700g. The endemic Pemba flying fox (*P. voeltzkowi*) is confined Pemba island in the Zanzibar archipelago, and is classified as Vulnerable, with a total population estimated at around 20,000–30,000 individuals, of which 90% are confined to just 10 roosting sites.

The largest species on the African mainland, the 450g hammerhead fruit bat (*Hypsignathus monstrosus*), is a colonial rooster associated with lowland rainforests in western Kenya and Uganda, where it can be distinguished by its bizarrely distended hollow head sac, which it uses as a loudspeaker to broadcast a booming mating call that can be heard from miles away on a calm night. Far more common and widespread, the straw-coloured fruit bat (*Eidolon helvum*) roosts in large, perpetually squeaking treetop colonies in suburban and rural Uganda – the colony resident in downtown Kampala's 'Bat Valley' moved on a few years ago, but you can still see large roosts in Jinja (near the golf course) and Entebbe.

The Egyptian fruit bat (*Rousettus aegyptiacus*) and its congeners are the only fruit bats to make use of a crude echolocation: they click their tongues rapidly and follow the echoes while confined in a cave or a darkened room, but under better-lit conditions revert to the more normal fruit-bat faculties of sight and sound. This species is also the only East African fruit bat that roosts in caves, as seen to spectacularly pungent effect in the bat caves at Maramagambo Forest in Queen Elizabeth National Park, where a million-strong colony provides a good source of food for a few resident pythons.

BIRDS

Fischer's lovebird (AVZ)

East Africa ranks as one the world's great ornithological destinations, with some 1,500 bird species recorded regionally, representing 15% of the global total. To place this in an African context, Kenya, Tanzania and Uganda respectively harbour the continent's second-, third- and fourth-greatest avian diversities: their checklists of roughly 1,160, 1,150 and 1,090 species are exceeded only by the 1,188 recorded in the Democratic Republic of Congo – a somewhat less tourist-friendly country that extends over an area larger than Kenya, Tanzania, Uganda and Rwanda combined!

For dedicated birdwatchers, a well-planned two-week itinerary through practically any part of East Africa is likely to result in a trip list of at least 350–400 species, a figure that compares favourably with anywhere in the world. The open savannah habitats of Kenya and Tanzania's popular game reserves form an ideal starting point for newcomers to African birdwatching – the likes of Serengeti, Maasai Mara, Samburu and Nyerere being just as rewarding for birds as for large mammals. For more experienced birders, however, East Africa's true gem is western Uganda, which harbours more than 100 forest species more generally associated with inaccessible parts of central or western Africa.

East Africa offers excellent birdwatching throughout the year, but the prime season runs from September and April, when the resident bird populations are boosted by those Palaearctic migrants from Europe and Asia that spend the northern hemisphere winter in equatorial or southern Africa. It has been estimated that as many as 6 billion individual birds undertake this trans-Sahara migration annually, ranging from the diminutive willow warbler and whinchat to the somewhat bulkier white stork, along with innumerable flocks of European (barn) swallow and various waders, wagtails, raptors and waterfowl. The European winter broadly coincides with East Africa's rainy season, when a number of resident birds, including weavers and their allies, undergo a startling 'ugly duckling'-style transformation from drab non-breeding plumage to gaudy breeding colours. This is also when most resident birds are singing and/or displaying, which makes them both easier to locate and identify, and more rewarding to watch.

WATERBIRDS

East Africa's rich variety of aquatic habitats embraces some 1,000km of Indian Ocean frontage as well as freshwater marshes, soda-crusted crater lakes, forest-fringed perennial rivers and a quintet of expansive 'inland seas' in the form of lakes Victoria, Tanganyika, Nyasa-Malawi, Turkana and Albert. Little wonder, then, that some 15% of the region's avifauna comprises waterbirds, ranging from widespread and familiar families such as gulls, ducks and waders to the more peculiar pelicans, flamingos and spoonbills.

Waterbirds are generally quite easy to observe and to identify. Most species forage openly around the water's edge or in the shallows, and tend to be less restless than their forest or savannah counterparts, allowing for long unobstructed views. The trickiest waterbirds to identify are probably seabirds such as skuas, petrels and various terns and gulls, which are usually seen in flight. Running them a close second are the waders, a group of largely drab shorebirds whose relatively sedentary habits do at least give you time to work on identification.

SEABIRDS

Seabirds are well represented off the shores of East Africa, with some 50 species of albatross, petrel, skua, frigatebird, shearwater, booby, tropicbird, gull, noddy and tern having been recorded. However, a significant majority of these are nomadic oceanic wanderers whose inclusion is based almost exclusively on sightings made at sea. Exceptions include a few resident and migrant gulls and terns, of which the

The grey-headed gull is the only gull regularly seen inland. (AVZ)

handsome lesser black-backed gull (*Larus fuscus*) is the most visible along the coast. The smaller and more nondescript grey-headed gull (*Chroicocephalus cirrocephalus*) occurs on inland waters, where it is often seen alongside the agile white-winged tern (*Chlidonias leucopterus*) and whiskered tern (*C. hybrida*).

AFRICAN SKIMMER

The eagerly sought African skimmer (*Rynchops flavirostris*), though somewhat tern-like in appearance, is significantly larger than any tern associated with African inland waters. Its black back and cap contrast strongly with the white throat and belly, but this bird's most striking feature is its long, laterally flattened red bill, whose lower mandible is almost 50% longer than the upper one. As its name suggests, the African skimmer feeds by flying low above the water, the lower mandible slicing the surface in search of food and upper mandible snapping shut when a suitable morsel is encountered. Small flocks are typically encountered on wide rivers and lakes fringed by the permanent sandbanks on which they roost and breed. Reliable sites include Lake Turkana, the Nile below Murchison Falls, and the Rufiji River in Nyerere National Park.

African skimmer in flight. (SS)

The great white pelican can be identified by its bright yellow pouch. (AVZ)

PELICANS

Two species of these larder-billed birds occur in East Africa. The great white pelican (*Pelecanus onocrotalus*), weighing in at 15kg, is the region's biggest waterbird and one of its most striking. Its predominantly white plumage is off-set by jet-black flight feathers (clearly visible in flight) and a large yellow pouch hanging from its massive bill. The slightly smaller and misleadingly named pink-backed pelican (*P. rufescens*) has a light grey back and dark grey flight feathers. You can often observe synchronised flotillas of ten or so 'great whites' bobbing on Rift Valley lakes such as Naivasha, Manyara and Baringo. The Kazinga Channel in Uganda's Queen Elizabeth National Park is a reliable spot for pink-backed pelican.

SHOEBILL

Topping the wish list of many dedicated African birdwatchers, the prehistoric-looking shoebill (*Balaeniceps rex*) is the proverbial odd bird: a slate-grey 150cm-tall marsh-dweller, whose enormous clog-shaped bill looks more something you'd expect to find in the pages of Dr Seuss than in a bona fide field guide! This outsized bill is generally fixed in a Cheshire-cat smirk that contrives to look both mildly sinister and cheerfully imbecilic, depending to some extent on whether it is viewed face-on or in profile; when agitated the bird claps it noisily, like a giant castanet.

Given its size, distinctive appearance, and relative abundance along the Sudanese Nile, it is surprising that the shoebill eluded Western taxonomists until the mid 19th century. In fact, early European explorers to the Sudan did return home with reports of a gigantic flying creature known to the local Arabs as *Abu Markub* (Father of the Shoe), but these were dismissed as pure fancy prior to 1850, when the eminent taxonomist John Gould chanced upon a specimen amongst a collection of birds shot on the Upper White Nile several years earlier. Describing his discovery as 'the most extraordinary bird I have seen', Gould placed it in a monotypic family and named it *Balaeniceps rex*: King Whale-head! Gould believed the strange bird to be most closely allied to pelicans, but it also shares anatomic characters with herons and until recently was held to be an offshoot of the stork family. Recent DNA studies support Gould's original theory that the shoebill is now thought to be most closely related to the hamerkop and both are placed in monospecific families of the order Pelecaniformes.

Confined to papyrus-dominated marshes in eastern equatorial Africa, the shoebill is an elusive and highly localised bird whose main population centre (up to 4,000 individuals, about half the global total) is the swampy Sudd, which flanks the Nile as it flows through southern Sudan. Within East Africa, it has been recorded as a vagrant in Kenya, while the Tanzanian population is more or less restricted to the Moyowosi–Kigosi Swamp in the little visited west of the country. In practical terms, Uganda is comfortably the best place to seek one, with half a dozen sites producing reliable sightings, notably the Nile Delta in Murchison Falls National Park, Toro-Semliki Wildlife Reserve, the Lake Albert Flats in Queen Elizabeth National Park and Mabamba Swamp west of Entebbe. The shoebill also occurs in small numbers in Rwanda's Akagera National Park, though sightings here are less frequent than in the more popular sites in Uganda.

LUNGFISH LOVER

The life cycle of the shoebill is just as remarkable as its appearance. It has an age span of 50 years and is generally monogamous, with pairs coming together during the breeding season to construct a grassy nest up to 3m wide on a small island or mound of floating vegetation. Two eggs are laid, and the parents rotate incubation duties, in hot weather filling their bills with water to spray over the eggs to keep them cool. The chicks hatch after about a month, and will be fed by the parents for at least another two months until their bills are fully developed. Usually only one chick per brood will survive. An adult shoebill consumes half its weight in food daily, preying on anything from toads to baby crocodiles, though lungfish are especially favoured. Its method of hunting is exceptionally sedentary: it might stand motionless for several hours before lunging with remarkable speed and power, wings stretched backward, to grab prey in its inescapable hook-tipped bill.

Shoebill (AVZ)

HERONS AND ALLIES

Arguably the most visible of East Africa's waterbird families, Ardeidae is also one of the most widely represented, with some 20 of the world's 65 species recorded. 'Typical' herons are long-necked, long-legged birds that stand 90–150cm high and use their sharp, elongated bills to spear fish, frogs and other prey. The grey heron

(*Ardea cinerea*) will be familiar to most European visitors and is commonly seen near water, while the less water-dependent black-headed heron (*A. melanocephala*) is often found foraging in cultivation and grassland. The giant of the group is the spectacular goliath heron (*A. goliath*), while the smallest and most elusive is the swamp-loving purple heron (*A. purpurea*), whose cryptic coloration and skulking habits make it hard to spot among reedbeds.

Egrets are similar in proportion to other herons but most have a predominantly white plumage – though some are uniform grey-black. Identification can be tricky, especially as three East African species all stand 65cm tall and come in both white and grey-black colour morphs, but bill and leg colour usually provide a clue. The 90cm great white egret (*A. alba*) can be distinguished by height alone, while the black heron (also known as black egret; *Egretta ardesiaca*) has the diagnostic habit of spreading its wings above its head to form a canopy that blocks the sun's reflection when it fishes – hence its nickname of 'umbrella bird'.

An atypical heron, the western cattle egret (*Bulbulcus ibis*) is by far the most common species in East Africa. This sociable, medium-sized, white bird often hangs around herds of cattle, buffalo or other grazers to prey on disturbed insects. At dusk, flocks of up to 50 are frequently seen flying along a watercourse to their communal roost.

The black-headed heron (*above,* AVZ) is often seen away from water. The much smaller black heron, or black egret *(left,* AVZ) has the unique habit of raising its wings to form an umbrella-like canopy while it fishes.

The western cattle egret feeds on insects disturbed by large mammals. (AVZ)

African darter (AVZ)

Nine other small heron species occur in East Africa, all of them rather squat and cryptically marked by comparison with egrets and typical herons. Most likely to be seen are the marsh-loving squacco heron (*Ardeola ralliodes*) and the widespread striated heron (*Butorides striata*).

CORMORANTS, DARTER AND FINFOOT

Cormorants are supreme fishers, distinguished by their long necks, strong bills and relatively stunted legs. They are poorly represented in East Africa, with just two of the world's 30-odd species present: the hefty and boldly marked white-breasted cormorant (*Phalacrocorax lucidus*) and more lightly built long-tailed cormorant (*Microcarbo africanus*). The latter is more widespread, but usually seen singly or in pairs, while the former is more gregarious and often seen in large flocks.

Fond of perching on bare overhanging branches, the African darter (*Anhinga rufa*) looks like an attenuated cormorant, with a kinked serpentine neck almost as long as its body and striking russet patches that glow golden in the right light. Its unusual habit of swimming with only its neck protruding above the surface lies behind the colloquial name of 'snakebird'.

The secretive African finfoot (*Podica senegalensis*), though not closely related to darters and cormorants, also has a long neck and tends to swim low in the water. It is among the most eagerly sought of African waterbirds, generally associated with quiet forest-fringed rivers and lakes, where it often swims below overhanging vegetation and is easily distinguished from similar species by its bright red bill. The most reliable site for finfoot in East Africa is Lake Mburo in Uganda.

STORKS

East Africa's eight species of stork are all tall, long-legged birds, superficially similar in appearance to herons, but with shorter necks, a stockier build and predominantly black-and-white (as opposed to grey-rufous) plumage. Migrant species, present from September to April, include the white stork (*Ciconia ciconia*), which will be familiar to many visitors from Europe, and the smaller black and Abdim's storks (*C. nigra*

and *C. abdimii*). Abdim's and white stork often form large mixed flocks that converge on recently burnt grass and insect emergences.

Of the resident stork species, the most common and widespread – and least water-dependent – is the marabou (*Leptoptilos crumenifer*), a fabulously ungainly omnivore that stands 150cm tall and can be distinguished by its scabrous bald head and inflatable flesh-coloured neck pouch. As befits a creature whose dark grey back and wings recall an undertaker's coat, the marabou is attracted to carrion and rubbish dumps, making it one of the most conspicuous birds in many East African cities (none more so than Kampala). It is also a common sight in many game reserves, often joining vultures to scavenge from kills.

Usually observed singly or in pairs in the vicinity of standing water, the saddle-billed stork (*Ephippiorhynchus senegalensis*) is a strikingly tall and handsome black-and-white bird rendered unmistakable by its red and black bill capped with a bright yellow saddle. Also associated with aquatic habitats, the yellow-billed stork (*Mycteria ibis*) resembles the European white stork but has a bright yellow (as opposed to red) bill and bold red face mask. The African openbill stork (*Anastomus lamelligerus*) is a glossy green-black ibis lookalike with a diagnostic gap between the mandibles of its long shell-cracker bill, while the woolly-necked stork (*Ciconia microscelis*) is a rather undistinguished black stork with a woolly-looking white neck and head.

IBISES AND SPOONBILLS

Ibises are robust birds with long decurved bills that probe the soil in search of snails, molluscs and other invertebrates. The most visible and audible species is the hadeda (*Bostrychia hagedash*), a grey-black bird, whose harsh onomatopoeic cackling is a characteristic sound of grassy wetlands and suburban hotel gardens. The less robust and much quieter glossy ibis (*Plegadis falcinellus*) is a glossy green-brown wetland specialist associated with marshes and lake shallows. Also very common in East Africa, the black-and-white African sacred ibis (*Threskiornis aethiopicus*) was revered in ancient Egypt (mummified birds have been

Above Marabou stork (AVZ)
Centre Saddle-billed stork (AVZ)
Below Yellow-billed stork (AVZ)

found in several Pharaonic tombs) but is extinct there today.

Closely related to the ibises, spoonbills are large all-white birds with distinctive long spatulate bills. The resident African spoonbill (*Platalea alba*) can be distinguished from the Eurasian spoonbill (*P. leucorodia*), a very scarce migrant, by its reddish (as opposed to black) face and legs. Spoonbills often forage in association with herons and egrets, sweeping their head from side to

Hadeda ibis (AVZ)

side while their partially submerged bill scythes through the shallows in search of small fish and invertebrates. The African Spoonbill also sometimes tags along behind wading crocodiles and hippos, latching onto any prey disturbed by their movements.

HAMERKOP

The rusty brown, medium-large hamerkop (*Scopus umbretta*) is endemic to sub-Saharan Africa, Arabia and Madagascar. It is one of the world's most singular bird species, placed in a monospecific family most closely related to the shoebill, and distinguished by a long flattened bill and angular crest that create the hammer-like appearance alluded to in its Afrikaans name (literally 'hammer head'). Likely to be seen in the vicinity of any well-wooded freshwater habitat, including temporary roadside ponds, the hamerkop is a diverse feeder, but its main prey consists of frogs and tadpoles, which it captures from riverside perches rather than foraging along the shore. It also sometimes follows buffalo and cattle, preying on disturbed grasshoppers and other insects.

In addition to its unusual appearance, the hamerkop is also known for its enormous nest, an untidy construction normally built in a tree fork close to the water. This is built over several months using branches, sticks, mud, litter, natural debris and pretty much anything else the builder can lay its beak on (human artefacts recorded in hamerkop nests include plastic bags, clothes lines, cardboard boxes, barbed wire and various items of clothing), and the finished product may measure 2m high and is similar in depth. The bird lays its eggs in a small chamber protected deep inside, suggesting that the purpose of this immense

Hamerkop (AVZ)

The lesser flamingo (*Phoeniconaias minor*) is the smaller of two species found in East Africa, but also the more pink. (AVZ)

construction is to create a stable microclimate for rearing a brood of fledglings. It has also been postulated that the large nest protects the bird and its young by attracting a veritable menagerie of squatters – but this seems a little improbable given that several of these usurpers are predators (snakes, monitor lizards, owls and small mammals) and would be more likely to eat the hamerkop chicks or force the parents out than to provide any protection.

FLAMINGOS

These distinctive pink-tinged birds have long exerted a fascination over humans, judging by the presence of 7,000-year-old cave paintings depicting flamingos in Mediterranean Europe and the ancient Egyptian use of the flamingo as a hieroglyph for the colour red. More contentiously, it has been suggested that mangled reports of flamingos breeding in the ash-strewn volcanic Natron area of northern Tanzania were the source of the ancient legend of the phoenix that rises from the ashes. Even if this is fanciful, there's no doubt that these colourful birds rank among the most popular avian attractions of the East African safari circuit, which supports an estimated 5–6 million greater and lesser flamingos (*Phoenicopterus roseus* and *P. minor*).

Often seen alongside each other in flocks of thousands, East Africa's two flamingo species both feed on algae and microscopic fauna, sifted through filters contained within their unique downturned bills. Flamingos are highly sensitive to the level and chemical composition of the shallow lakes wherein they feed, and will readily relocate when conditions change, which means that a site that hosts immense flocks one month might be abandoned the next. Among the more reliable sites in East Africa are lakes Bogoria and Nakuru (up to 2 million birds present in perfect conditions), Lake Manyara and the Ngorongoro Crater in northern Tanzania, and certain crater lakes in Uganda's Queen Elizabeth National Park.

CRANES

Superficially similar to storks but more closely related to rails and bustards, cranes are striking metre-high birds associated with marshy and grassland habitats. The only widespread East African resident is the grey crowned crane (*Balearica regulorum*), an astonishingly beautiful grey, white and chestnut bird distinguished by its bristly golden crown and red neck wattle. The national bird of Uganda, the grey crowned crane is common in moist wetlands throughout the south and west of that country, and is also likely to be observed on safaris to the relatively moist Mara-Serengeti-Ngorongoro ecosystem of northern Tanzania and southern Kenya. It is replaced in the far north of Uganda and the Turkana region by the similar but arguably even more handsome black crowned crane (*B. pavonina*). The East African range of the endangered wattled crane (*Grus carunculata*) is restricted to the Moyowosi–Kigosi Swamp and a few other remote localities in southern Tanzania.

The grey crowned crane is the national bird of Uganda. (AVZ)

The region's most common waterfowl is the assertively territorial Egyptian goose. (AVZ)

DUCKS, GEESE AND GREBES

The freshwater and saline lakes of East Africa support 14 resident and 10 migrant species of waterfowl, all of which are placed in the family Anatidae. One of the most conspicuous is the ubiquitous Egyptian goose (*Alopochen aegyptiaca*), a large rufous-brown waterfowl that probably wasn't named in reference to Cairo's legendarily obstreperous taxi drivers, though its assertive demeanour and perpetual honking might suggest otherwise. The spur-winged goose (*Plectropterus gambensis*) is East Africa's bulkiest waterfowl, weighing in at almost 10kg, while the knob-billed duck (*Sarkidiornis melanotos*) is notable for the large comb on the breeding male's bill.

The aptly named white-faced whistling-duck (*Dendrocygna viduata*) is often seen in large flocks alongside the closely affiliated but less numerous rufous whistling-duck (*D. bicolour*), while the less sociable but equally well-named yellow-billed duck (*Anas undulata*) is common on highland lakes and streams. The region's smallest and most

Yellow-billed duck (AVZ)

beautiful waterfowl is the African pygmy goose (*Nettapus auritus*), a green-backed, chestnut-bellied and yellow-billed resident of still lily-covered waters such as Mabamba Swamp in Uganda. Good sites for those who want to tick a wide variety of resident and migrant ducks (the latter typically present October–March) include Arusha National Park, the Ngorongoro Crater in the rainy season, and most of the smaller Rift Valley lakes of southern Kenya and northern Tanzania.

Although unrelated to ducks, grebes (family Podicipedidae) bear a strong resemblance in general body shape, but have sharper arrow-like bills and are seldom seen out of the water. The little grebe or dabchick (*Tachybaptus ruficollis*), is probably the most common duck-like bird in the region, and is significantly smaller than any African duck. More distinguished-looking but relatively uncommon, the black-necked grebe (*Podiceps nigricollis*) and striking great crested grebe (*P. cristatus*) are most likely to be observed on the smaller Rift Valley lakes of Kenya and Tanzania.

RAILS AND ALLIES

Rallidae is a diverse family of small to medium-sized water and/or forest-associated birds whose general demeanour recalls an unusually dumpy fowl. Most species are characterised by short legs, long toes, small rounded wings, and short but perkily upturned tails. Although 20 species are known from East Africa, most are highly secretive and unlikely to be seen unless actively sought. Exceptions include those rails associated with open water, such as the common moorhen (*Gallinula chloropus*) and highly gregarious red-knobbed coot (*Fulica cristata*), both of which swim in open water like ducks or grebes. A close scan of lush marginal vegetation will often reveal the presence of two other rails: the petite and unusually confiding black crake (*Zapornia flavirostra*) and the bright purple, green and red African swamphen (*Porphyrio madagascariensis*).

WADERS

East Africa's checklist includes roughly 35 species of Scolopacidae, a diverse family of small to medium-sized sandpipers. Many of these waders are listed on the basis of a handful of vagrant records; some form part of the regular influx of many thousands of Palaearctic migrants during the European winter; and others represent a scattered resident population, boosted by significant numbers of seasonal migrants. In their African non-breeding plumage, most waders are nondescript grey, brown and/or white shorebirds with a slender neck, a long bill, a restless demeanour and few strong field characters.

Identification can pose a serious challenge to inexperienced observers and requires a good field guide. When viewing an unfamiliar wader species, make a note of the approximate size, the length and shape of the bill (straight, decurved or upcurved), the leg colour and any notable markings – including the prominence (or lack) of an eye stripe. Wader identification is easiest during the European summer, when the inland possibilities are more or less restricted to a handful of species, some individuals of which remain in East Africa all year round. These include the common sandpiper (*Actitis hypoleucos*), wood sandpiper (*Tringa glareola*) and common greenshank (*T. nebularia*).

The purple swamphen (*above*, H/D) is a colourful but secretive reedbed resident, whereas the marsh sandpiper (*below*, AVZ), like most waders, frequents open shallows.

Also widespread are two striking and easily identified members of *Recurvirostridae*, a family of birds that, despite their sandpiper-like appearance, are actually more closely related to the plovers. The black-winged stilt (*Himantopus himantopus*) is a slender black-and-white bird with a straight, needle-like black bill and very long red legs, associated with a wide range of shallows, from the Rift Valley lakes to coastal lagoons. More handsome still (and a firm favourite of many European birders) is the pied avocet (*Recurvirostra avosetta*), a tall black-and-white shorebird with a distinctive upward-curving bill. It is uncommon on the coast and far west of the region, but often occurs in significant numbers in the smaller Rift Valley lakes of Kenya and Tanzania.

PLOVERS, COURSERS AND THICK-KNEES

Superficially similar to sandpipers, but with longer legs, shorter bills, a bulkier build, greater variation in colour and more distinct field characters, plovers are well represented in East Africa, with some 24 species present. Particularly conspicuous are the 11 boldly marked and relatively tall lapwings of the genus *Vanellus*, most of which are extremely vociferous and will mob intruders excitedly. Several species are strongly associated with water: the long-toed lapwing (*V. crassirostris*) favours swampy habitats with floating vegetation, while the white-crowned lapwing (*V. albiceps*) is restricted to wide riverine sandbars such as those along

The black-winged stilt (*above*) has proportionally the longest legs of any wader. It is often seen alongside the spur-winged lapwing (*centre*) in the south of the region and less common black-headed lapwing (*below*) in the northwest. (AVZ)

Kittlitz's plover is usually seen alone or in pairs. (AVZ)

the Rufiji in southern Tanzania. You might observe the more widespread blacksmith and spur-winged lapwings (*V. armatus* and *V. spinosus*) alongside practically any wetland habitat. Of the grassland plovers, the crowned lapwing (*V. coronatus*) is one of the more visible roadside birds in many game reserves, though it is replaced by the black-headed lapwing (*V. tectus*), with its striking spiked crest, in more arid parts of northern Kenya and Uganda. Freshly burnt grassland often attracts the nomadic Senegal and brown-chested lapwings (*V. lugubris* and *V. superciliosus*).

The smaller 'sand plovers' of the genera *Charadrius* and *Anarhynchus* are represented by four resident and seven migrant species, most of which are strongly associated with water and must be identified with care. The most widespread and common are the handsome three-banded plover (*C. tricollaris*) and buff Kittlitz's plover (*A. pecuarius*), either of which might be observed singly or in pairs foraging busily alongside any river, lake or wetland.

Similar in build to the *Vanellus* plovers but a little smaller, and far less common and visible, the coursers are crepuscular, fast-running ground birds represented by six species in East Africa, some of which are associated with waterside habitats and others with arid or recently burnt grassland. The superficially similar Egyptian plover (*Pluvianus aegyptius*) is a striking grey, black, buff and white bird, mainly associated with sandy riverine habitats in the Sahel, but also sometimes observed in the far north of Uganda. Also rather plover-like in appearance, but larger and stockier, the thick-knees (also known as dikkops – thickheads!) are well camouflaged brownish birds with bright yellow legs, eyes and beak bases. Of the four species recorded in East Africa, the most easily identified is the mostly nocturnal spotted thick-knee (*Burhinus capensis*), which has black-on-brown feathering and inhabits a variety of savannah habitats. The other three are restricted to waterside habitats and are practically indistinguishable in the field.

The spotted thick-knee tends to rest up by day, but is often conspicuously vocal at dusk. (AVZ)

The African jacana is often observed walking delicately over floating vegetation. (AVZ)

JACANAS AND PAINTED SNIPES

Also known as lily-trotters, the jacanas are singular waterbirds notable for the exceptionally spreading toes that allow them to walk on lily pads and other light floating vegetation in marshy or well-vegetated pools and river edges. The African jacana (*Actophilornis africanus*) is one of the region's trademark waterbirds, and is totally unmistakable with its rich chestnut torso and wings, white neck, black cap and blue bill and frontal shield. By contrast, the lesser jacana (*Microparra capensis*) is rather localised and uncommon, and can be difficult to identify positively due to its strong resemblance to the immature African jacana. Closely related to the jacanas but placed in a separate family represented by just two species worldwide, the greater painted-snipe (*Rostratula benghalensis*) is a distinctive and rather beautiful snipe-like shorebird, whose cryptically marked wings contrast with its white breast, orange head and bold white eye-ring. Unusually, when you see an adult painted snipe or African jacana with a youngster, you can assume that the adult is male. These two species fall among the 1% of birds that practise polyandry, the name given to a courtship system wherein one female will mate with several males. This system is generally twinned with male-only parental care.

RAPTORS

Raptors are arguably the most charismatic of birds and have held humankind enthralled since prehistoric times. With their strong hooked bills, sharp talons, keen eyesight and predatory prowess, they are the feathered equivalent of the lions and other mammalian carnivores that stalk the African savannah. Even today their appeal extends beyond hardcore birdwatchers to most casual safari-goers.

East Africa boasts a mind-boggling variety of raptors. These were traditionally all placed in the order Falconiformes, but DNA tests undertaken in 2008 revealed that falcons and allies are more closely related to passerines and near-passerines than to hawks and eagles. As a result, raptors are now split between two orders: Accipitriformes (which includes hawks, eagles and Old World vultures) and Falconiformes (falcons, caracaras and similar smallish raptors). Within the former, Accipitridae is East Africa's most numerically diverse bird family, with more than 60 species recorded. There were also two monotypic Accipitriformes families, namely Sagittariidae (secretary bird) and Pandionidae (osprey), while Falconidae is represented by 19 species. All in all, it's a mouthwatering destination for raptor enthusiasts, though identification can often be challenging, so you will certainly need a decent field guide.

The tawny eagle is not above scavenging from carcasses. (AVZ)

AQUILA EAGLES

The true eagles of the genera *Aquila*, *Hieraaetus* and *Clanga* are the archetypal birds of prey – swift and powerful in flight, literally eagle-eyed, and impressively big when perched, though some species are very difficult to distinguish from one another in the field. Five of the world's 19 species are resident in East Africa, another three are regular Palaearctic migrants, and two (imperial and greater spotted eagle: *A. heliaca* and *C. clanga*) have been recorded a couple of dozen times only.

Largest and most impressive is the spectacular Verreaux's eagle (*A. verreauxii*), which weighs up to 5kg and has a wingspan of 2.5m. Mainly black in colour – and also known as 'black eagle' – this massive cliff-nesting species has a bold yellow beak and legs, a diagnostic combination of white rump and white 'V' marking on its back, and pale bases to the primary wing feathers that form distinctive 'windows' in flight. It is normally observed singly or in pairs, soaring effortlessly around rocky mountains and cliffs, occasionally with a paler immature bird in tow. Fortunate observers might witness the male's aerial courtship display, during which it performs spectacular cartwheels, rising or falling hundreds of metres in one powerful swoop, accompanied by a far-carrying high pitched screech. A swift and efficient hunter, Verreaux's eagle feeds on anything from baboons and young gazelles to geese and herons, but it is especially partial to hyraxes, which share its rocky habitat. Good sites include Hell's Gate and the cliffs behind Lake Baringo in Kenya, and the Serengeti's Lobo Hills and the Rift Valley escarpment above Lake Manyara in Tanzania.

The eastern (or southern) banded snake eagle typically hunts reptiles from treetop perches in the forests of the coastal region. (AVZ)

In East Africa, the most common member of the genus is the tawny eagle *A. rapax*, the region's largest all-brown resident raptor, with a 2m wingspan. The adult has a uniform brown plumage, but the tone varies greatly between individuals, ranging from dirty blond through tawny-buff to dark brown. It is generally paler than the very similar steppe eagle (*A. nipalensis*), a non-breeding migrant that is widespread between October and April. The best way to distinguish these two species is to look at the bright yellow gape, which extends behind the eye in the case of the steppe but only halfway below the eye in the case of the tawny. Both species are agile hunters with a varied diet, but the tawny eagle subsists mainly on small mammals and carrion, whereas the more insectivorous steppe eagle often congregates at termite and other insect emergences.

The smaller Wahlberg's eagle (*H. wahlbergi*), though scarce between May and July, is quite common during the rest of the year, when the resident population is boosted by an influx of intra-African migrants. Most individuals are dark brown, but others may be off-white or intermediate in colour. In all cases, Wahlberg's eagle is best distinguished from other brown eagles, such as the relatively uncommon lesser spotted eagle (*C. pomarina*), by its small occipital crest and distinctive narrow, square-tipped tail in flight.

The African hawk-eagle (*A. spilogaster*) and Ayres's hawk-eagle (*H. ayresii*) are streamlined medium-sized eagles with black upperparts, black-on-white streaked underparts and yellow feet. Neither is particularly common: the larger and less densely streaked African hawk-eagle might be observed in practically any wooded habitat, often hunting in pairs, whereas Ayres's hawk-eagle prefers riparian and other moist forests.

SNAKE-EAGLES

The snake-eagles of the genus *Circaetus* are represented in East Africa by four resident species. These medium-large raptors have a distinctive upright stance, and their combination of large rounded head and piercing pale eyes is almost owl-like. All species tend to be solitary and to perch conspicuously on a bare branch near the top of a tree, from where they hawk their prey. As their name suggests, snake-eagles feed to a greater or lesser extent on snakes, though lizards, small birds and mammals are also taken. Many feed extensively on the likes of mambas and cobras, despite having no immunity to their venom, which means that the slightest misjudgement in an attack might prove fatal.

The brown snake-eagle (*C. cinereus*) is a conspicuous resident of savannah and woodland habitats, and can be distinguished from other large all-brown raptors by the combination of yellow eyes and no yellow elswhere. Equally widespread but somewhat less common, the black-chested snake-eagle (*C. pectoralis*) is a handsome white-bellied bird with a black back, head, wings and chest. The western and eastern banded snake eagles (*C. cinerascens* and *fasciolatus*) are dark grey-brown forest-edge birds with barred breasts, and would be difficult to tell apart were their ranges not mutually exclusive.

FISH EAGLE AND OTHER WATERSIDE RAPTORS

Three of East Africa's larger raptors are almost always seen perched near water. Most common, indeed the first bird that many neophyte safari-goers learn to identify, is the African fish eagle (*Icthyophaga vocifer*) a magnificent black-and-white eagle with a rich chestnut belly and yellow base to its large hooked bill. Strongly monogamous, the African fish eagle is conspicuous along the shore of most rivers and lakes, where it is as likely to be seen perched in a large waterside tree as it is in soaring flight – occasionally sweeping down to the surface to scoop a fish into its large talons. As alluded to in its scientific name, it has a far-carrying and ringing call, which is often delivered as a duet with both birds throwing back their heads dramatically. This is one of the most evocative sounds of the African bush: once heard, never forgotten!

As the name suggests, African fish eagles primarily eat fish. (AVZ)

The palm-nut vulture might resemble the fish eagle and share its habitat, but the shape of its bill reveals its vulturine affinities. (AVZ)

The palm-nut vulture (*Gypohierax angolensis*), though not closely related to the African fish eagle, is similar enough in appearance to have earned the alternative name of vulturine fish eagle. But it has much more white in its plumage, no chestnut on its belly, a longer and squarer bill, and a diagnostic red mask stretching behind its eyes. It does not feed on fish, and its association with water is due to the presence of the raffia palms whose fruit form the core of its diet (supplemented by carrion, small mammals and various invertebrates). The distribution of the palm-nut vulture mirrors that of the raffia palm: it is common along parts of the Lake Victoria shore (particularly Entebbe) and the Indian Ocean coastline, as well as raffia-lined rivers such as the Rufiji in southern Tanzania.

More familiar to European visitors, the osprey (*Pandion haliaetus*) is a thinly distributed resident whose numbers are significantly boosted by Palaearctic migrants between October and March. It is placed in a separate family to other Accipiters on the basis of the unique overlapping scales that cover its legs. It is also among the most widely distributed birds, with a range extending to all non-polar corners of the globe. This highly specialised species feeds almost entirely on fish, both in marine and freshwater environments, typically hunting shortly after dawn or before dusk, circling above the water until it picks out a likely meal, then plunging to the surface with talons stretched downward. It is quieter and more solitary than the African fish eagle, and might be seen perched above any watery habitat – especially well-wooded freshwater lakes.

OTHER EAGLES

The spectacular bateleur (*Terathopius ecaudatus*), placed in a monospecific genus, derives its common name from the French word for tightrope walker, which well describes its unique tilting flight, as though wobbling along a suspended line. Often seen in pairs, this monogamous bird is rendered unmistakable, when perched, by its stocky build and unique plumage: black with an orange (or occasionally white) back, and bright red face mask and legs. It is also very distinctive in flight, thanks to the combination of its unusually wide, narrow-tipped wings (white with a black margin when seen from below), stunted orange tail and trademark tilting motion. You can see one sweeping through the skies in any East African savannah reserve, where it feeds largely on carrion, though it is also an adept hunter of small mammals, birds and reptiles.

The bateleur's striking markings and tilting flight make it one of the most distinctive of all large raptors. (AVZ)

The martial eagle is one of the region's largest and most powerful raptors. (AVZ)

Another spectacular resident of savannah and woodland, the martial eagle (*Polemaetus bellicosus*) is a massive and powerfully built raptor with a brown-black back, head and chest, a small but distinct occipital crest, and a white breast lightly speckled with black – though immature birds are much whiter in appearance. Reaching weights of over 5kg, the martial eagle is capable of downing antelope that weigh twice as much, though it mostly feeds on game birds and small mammals. Largely confined to recognised conservation areas, this species is widespread in East Africa and one of the most familiar raptors on safari thanks to its unmistakable appearance, great size and preference for open perches.

Similar in size and body shape, the crowned eagle (*Stephanoaetus coronatus*) is a bulky chestnut-brown/grey raptor with a rough occipital crest and densely barred underparts. It is confined to dense woodland and forest habitats, where it uses its massive talons to capture smallish mammals such as hyraxes and monkeys. In much of East Africa this is the only forest-dwelling eagle, though in western Uganda you might confuse it with the smaller Cassin's hawk-eagle (*Aquila africana*) or Congo serpent eagle (*Dryotriorchis spectabilis*).

The unmistakable long-crested eagle (*Lophaetus occipitalis*) is a medium-large dark brown raptor with bold yellow eyes, feet and cere, and a long windblown occipital crest that gives it a rather foppish appearance. Often seen perching openly on treetops or telegraph poles, from where it pounces on small mammals, it is probably the commonest eagle in moist parts of East Africa such as the well-wooded grassy highlands of central Kenya, northern Tanzania and western Uganda.

BUZZARDS

Buzzards are powerful, medium-large raptors with long broad wings, relatively short tails and a stocky eagle-like appearance when perched. Only two species are likely to be encountered by casual visitors: the resident augur buzzard (*Buteo augur*) and migrant steppe buzzard (*B. buteo*). The former is a conspicuous and handsome inhabitant of moist highlands (eg: north of Nairobi and in the vicinity of Ngorongoro), with a black back, white (or sometimes black) breast, and an orange-red tail that precludes confusion with any raptor other than the bateleur. By contrast, the steppe or common buzzard is a nondescript brownish bird with faintly barred underparts and in most cases an indistinct (but, where present, diagnostic) pale band between the breast and the belly. It is common in October and March, when Palaearctic migrants heading to and from southern Africa pass through, but reasonable numbers are present throughout the northern winter.

KITES

The black kite (*Milvus migrans*) is probably the most common raptor in East Africa, particularly during the northern hemisphere winter when migrants join the resident yellow-billed kite (*M. aegyptius*), which is sometimes treated as a subspecies of black kite. A medium-large, dark-brown, fork-tailed raptor with a distinctive tail-twisting flight, it is unusually urbanised, with large flocks frequently seen circling and swooping gracefully above towns and villages. It can also be a bold scavenger, as picnickers in the Ngorongoro Crater often find to their cost. The other common kite in East Africa, the smaller and less sociable black-winged kite (*Elanus caeruleus*) looks rather plump and dove-like when perched, but is unmistakable in flight, often hovering almost motionless for several minutes before swooping down onto a rodent or small bird.

The black kite (*above*, AVZ) is often seen circling above towns and villages, while the smaller black-winged kite (*top*, AVZ) is easily identified by its habit of hovering.

HARRIERS

These elegant medium-sized raptors spend much of their time flying low above grassland or wetland habitats, quartering the ground for prey, and the four East African species can be difficult to identify with certainty. The pallid and

Eastern chanting goshawk (AVZ)

Montagu's harriers (*Circus macrourus* and *C. pygargus*) are both silver-grey (male) or brown (female) in general coloration, and frequent open grasslands such as the Serengeti Plains, where they are conspicuous during the northern hemisphere winter. By contrast, the much browner African marsh and western marsh harriers (*C. ranivorus* and *C. aeruginosus*) are found almost entirely on wetlands and flooded grasslands, with the former being present throughout the year and the latter only from October to April.

SPARROWHAWKS, GOSHAWKS AND ALLIES

Represented by ten species in East Africa, the sparrowhawks and goshawks of the genus *Accipiter* are small to medium-large raptors, typically characterised by plain dark grey or brown upperparts, paler barred underparts, short broad wings and long barred tails – though several species also come in a rare melanistic (all black) morph. The most common species are the shikra (*A. badius*), a dove-sized raptor of savannah and light woodland; the African goshawk (*A. tachiro*), a medium-sized, densely barred hawk that's most often glimpsed flying swiftly between the trees; and the little sparrowhawk (*A. minullus*), which is smaller even than the shikra and associated with denser woodland. Less common but also resident, the impressive great (or black) sparrowhawk (*A. melanoleucus*) is a buzzard-sized, boldly marked black-and-white bird associated with indigenous and exotic forest, and dense woodland.

Several other East African raptors are loosely allied to the sparrowhawks. Most conspicuous perhaps are the eastern and western chanting goshawks (*Melierax poliopterus* and *M. metabates*), buzzard-sized raptors with distinctive upright stances, grey backs and finely barred underparts, pale rumps and red legs. The eastern species is paler than its western counterpart and has a yellow (as opposed to red) bill, and their ranges are mutually exclusive. The gabar goshawk (*Micronisus gabar*) is similar in most respects but much smaller, while the misleadingly named lizard buzzard (*Kaupifalco monogrammicus*) can be distinguished from all other raptors by the bold black vertical stripe on its white throat.

A fairly common and conspicuous resident of wooded habitats, the African harrier-hawk or gymnogene (*Polyboroides typus*) is a large grey nest-raider with a slim head, yellow mask, finely barred underparts and long legs with flexible joints, adapted to climbing trees in search of eggs and chicks. Another unusual hawk-like bird placed in a monotypic genus is the crepuscular bat hawk (*Macheiramphus alcinus*), which preys on bats, and is almost always seen around caves and trees in which they roost. In flight this species has more pointed, falcon-like wings.

VULTURES

It might be stretching a point to describe vultures as charismatic, but the sight of these gigantic scavenging raptors massing around a fresh carcass, squabbling and squawking and poking at each other, is certainly among the most memorable that a safari has to offer. And while a flock of feeding vultures seldom displays any great dignity, to burden these ecologically important birds with negative anthropomorphic associations is to ignore the vital role they play in clearing away the carcasses that would otherwise litter the African bush. Furthermore, if one needs to place a few extra points on the merit sheet, vultures are among the most efficient fliers in the avian kingdom, capable of soaring on thermals for hours on end, their vision is practically unmatched in the animal kingdom, and – contrary to popular perception – they are fastidiously clean, and spend long periods preening themselves to perfection after a messy meal.

It's not unusual to see most or all of East Africa's six carrion-eating vulture species gathered on the same carcass, nor is it difficult to tell them apart. Africa's largest raptor, the lappet-faced vulture (*Torgus tracheliotus*) is a truly massive bird, typically seen singly or in pairs, with a rather square bald pink head, a blue lower face, and

White-backed vulture (AVZ)

The lappet-faced vulture is the largest African raptor. (AVZ)

heavy black wings that it spreads open like a cape, reinforcing its rather ghoulish presence. The slightly smaller white-backed and Rüppell's griffon vultures (*Gyps africanus* and *G. rueppelli*) are still larger than most eagles, and often comprise 90% of the vulture headcount at a kill. The easiest way to tell these similar species apart is that the adult Rüppell's griffon has a scaly pattern to the upperparts and a horn-coloured (as opposed to black) bill.

The slightly smaller white-headed vulture (*Trigonoceps occipitalis*), though relatively uncommon, is easily distinguished from all other species by its white crest and heavy red bill. The only vulture regularly observed near human habitation is the scruffier and even smaller hooded vulture (*Necrosyrtes monachus*), an all-brown bird with a bald red face and much narrower bill. The distinctive but sparsely distributed Egyptian vulture (*Neophron percnopterus*) is a medium-sized off-white raptor with a shaggy white crest, yellow face mask and black-tipped bill. Although it might turn up anywhere in the region, it is most common in the arid badlands of northern Kenya, where it is often seen near water. This species is one of the few known avian tool-users, having been observed using a stone to crack open an ostrich egg.

SECRETARY BIRD

Placed in a monotypic family ancestral to all other modern members of Accipitriformes, the secretary bird (*Sagittarius serpentarius*) is a bizarre raptor. It stands up to 1.5m tall and is unmistakable with its long skinny legs, grey plumage, long black wings and tail, and bare red face mask. It is often said that the bird is named for its swept-back black crest, which vaguely recalls the quills used by Victorian secretaries, but it is more likely that 'secretary' is a corruption of the Arabic *saqr-et-tair* (hunting bird). Particularly common in the southern Serengeti, single birds and pairs might be seen striding purposefully though any grassland or light wooded savannah in search of their favoured prey of snakes, which they stamp to death in a flailing dance ritual that might just be the African equivalent of the funky chicken! Although it is a terrestrial hunter, the secretary bird does roost in trees by night, and it will take to the air when disturbed – its long take-off is reminiscent of an aeroplane on a runway, though its high-soaring flight is magnificent.

FALCONS

East Africa is home to 19 species of falcons, a family of small to medium-sized raptors characterised by their slim build and fast direct flight. Most have cryptic and subtly variable markings that can make identification quite a challenge. All but one of the East African species are assigned to the genus *Falco*, the exception being the peculiar pygmy falcon (*Polihierax semitorquatus*), a pretty, dry-country raptor so diminutive that it is often mistaken for a shrike. Most other falcon species are uncommon or inconspicuous. Among those likely to be observed by casual safari-goers are the common kestrel (*F. tinnunculus*), which is speckled brown with grey cap and yellow eye-ring; the very similar but slightly pointier-tailed lesser kestrel (*F. naumanni*); the grey kestrel (*F. ardosiaceus*), which is all grey with a large yellow eye-ring; the dashing lanner falcon (*F. biarmicus*), which has a grey-brown back, rust-washed white underparts, a black moustache and red cap; and the less common but equally splendid peregrine falcon (*F. peregrinus*) which has a grey-brown back, barred underparts, a black moustache and cap, and has been reliably recorded as the fastest bird in the sky.

The secretary bird (*above*, ss) and greater kestrel (*Falco rupicoloides; below*, avz) are two very different raptors both associated with open country.

Two male ostriches fighting. (AVZ)

GROUND BIRDS
OSTRICH

Despite being the world's largest birds, standing 2m tall and weighing in at 100kg, ostriches are relatively miniaturised relics of a formerly more diverse order of hefty flightless birds known as ratites. Absolutely unmistakable, ostriches display strong sexual dimorphism, with the male being larger and very handsome with its striking black and white plumage, and the female being smaller, duller and scruffier. Two ostrich species are recognised: the common ostrich (*Struthio camelus*) of Tanzania and west-central Kenya, which has a pink neck and legs, and the Somali ostrich (*S. molybdophanes*) of northeast Kenya, whose neck and legs are blue-grey. Known for its outsized eggs, the ostrich is increasingly popular as a source of meat, but unlike in southern Africa, ostrich farming is not widely practised in East Africa. In the wild it is associated mainly with short grassland and arid habitats, where it remains quite common.

BUSTARDS

Loosely related to cranes, the bustards of the family Otididae are sturdily built medium-to-large ground birds associated with open grassland, desert and lightly wooded savannah. The most conspicuous East African species is the enormous kori bustard (*Ardeotis kori*), which stands up to 1.3m tall and weighs up to 12.5kg, making it a contender for the title of the world's heaviest flying bird. This species is a widespread resident of grassland habitats in Kenya and northern Tanzania, and especially common in the Serengeti-Mara complex. Seven smaller bustard species have also been recorded in the region, with the black-bellied and white-bellied bustards (*Lissotis melanogaster* and *Eupodotis senegalensis*) being the most visible species on the popular safari circuits of northern Tanzania and southern Kenya.

Hartlaub's bustard (*Lissotis hartlaubii*; *above*) is one of several similar medium-sized bustards in the region, but none compares in size to the massive kori bustard (*right*). (AVZ)

157

The grey-breasted spurfowl (*above*) is endemic to the Serengeti, where it is very common. The helmeted guineafowl (*below*), by contrast, is ubiquitous throughout East Africa. (AVZ)

FRANCOLINS AND OTHER FOWL

Also known as game birds, a term with archaic hunting associations, the chicken-like birds in the order Galliformes are the most characteristic of East Africa's ground birds. Some 28 species are recognised in the region, split between Phasianidae (francolins) and Numididae (guineafowls). Phasianidae is a group of plump and mostly grey or brown birds, generally similar in size and behaviour to a domestic hen. Some are rather drab and difficult to tell apart in the field, while others are very distinctive, most notably perhaps the yellow-necked spurfowl (*Pternistis leucoscepus*) of Kenya and northeast Tanzania. One species, the grey-breasted spurfowl (*P. rufopictus*), is endemic to Tanzania, its range centred on the southern Serengeti. Also endemic to Tanzania, the remarkable Udzungwa forest-partridge (*Xenoperdix udzungwensis*) of the Udzungwa Mountains was first discovered for Western science as recently as 1991, and is more closely related to the Asian hill-partridges than to any other African bird.

Far more striking and even noisier than the francolins are the guineafowls, an endemic African family of large, gregarious fowls, represented in the region by three species, all of which are spotted white on grey and have bare blue heads. The ubiquitous helmeted guineafowl (*Numida meleagris*) inhabits a wide variety of habitats and has a completely bare head with an ivory coloured helmet. The striking crested guineafowl (*Guttera pucherani*), a relatively localised inhabitant of riparian woodland and forest interiors, is similar in appearance but can be distinguished by its shaggy mop-top crest and, in the case of the northern races, a red face or neck mask. The star of the family, however, is the vulturine guineafowl (*Acryllium vulturinum*), which has a brilliant cobalt chest covered in lacy black-and-white feathers, and is associated with dry-country habitats in northern Kenya, especially Samburu-Buffalo Springs.

NEAR-PASSERINES

The term near-passerines is applied collectively to 11 bird orders that lack the distinctive toe pattern common to all passerines (page 174) but otherwise resemble this most diverse of avian orders in general appearance and ecology. The taxonomic validity of the term is debatable, to say the least, but it remains a convenient catch-all phrase to describe a disparate group of mostly arboreal (and in many cases brightly coloured) birds, including doves, parrots, cuckoos, owls, nightjars, swifts, kingfishers, rollers, woodpeckers and barbets.

DOVES AND PIGEONS

The plump, small to medium-sized birds of the family Columbidae need little introduction, with several of the world's most visible and vocal species being classed among their number. Common representatives include the ring-necked and red-eyed doves (*Streptopelia capicola* and *S. semitorquata*), a pair of greyish ring-necked birds whose repetitive calls – respectively transliterated as 'work harder, work harder' and 'I am a red-eyed dove' – are among the definitive sounds of the African bush, as is the trademark five- or six-syllable cooing call of the smaller laughing dove (*S. senegalensis*), which differs from the first two by its lack of collar.

Also very commonly seen on game drives in woodland habitats is the emerald-spotted wood-dove (*Turtur chalcospilos*). This pocket-sized dove is the most widespread member of a genus whose four East African representatives all have brown wings with iridescent blue or green wing spots and a lovely peaceful call comprising up to 15 soft descending notes. Typically associated with drier and more open habitats than most other columbids, the even smaller Namaqua dove (*Oena capensis*) is a tiny dove with a long graduating tail and, in the case of the male, a distinctive black face and chest.

Pigeons are larger than doves and, in East Africa, tend be associated with more forested habitats – though the indigenous speckled (or rock) pigeon (*Columba guinea*), with its distinctive red eye-mask, is most likely to be seen around cliffs, and also adapts well to urban settings, where it is outnumbered by the introduced feral pigeon (*C. livia*). The African green pigeon (*Treron calvus*) is the most colourful

Above Speckled pigeon (AVZ)
Below Emerald-spotted wood-dove (AVZ)

of the region's columbids. This bright green fruit-eater is often seen in riparian forest, especially around fruiting fig trees. It is replaced in the far north by the yellow-breasted Bruce's green pigeon (*T. waalia*) and on Pemba by the grey-breasted Pemba green pigeon (*T. pembaensis*), which is endemic to that island.

SANDGROUSE

East Africa's five sandgrouse species (family Pteroclidae) are closely related to pigeons and similar in size and appearance. They are ground birds of open grasslands and arid country, and are most active in the early morning and late evening, when large flocks tend to congregate at pools and lakes. You might also see them at other times of day scurrying singly or in pairs alongside game reserve roads. Most species are brown in general coloration, with pretty though cryptic patterns that allow for easy identification, and many have rather peculiar and memorable calls. The most common species in the popular game reserves of Kenya and northern Tanzania are the black-faced,

Black-faced sandgrouse (AVZ)

chestnut-bellied and yellow-throated sandgrouse (*Pterocles decoratus, P. exustus* and *P. gutturalis*). The daily activities of these arid-country specialists revolve around the need to drink once every 24 hours. Adult birds will routinely fly up to 60km daily in either direction to visit the nearest water source, where they drink very quickly, continuously sucking up water through their submerged beaks until they are full. Sandgrouse are the only birds that can carry water by letting it soak into their abdominal feathers. An adult can soak up about 20ml of water to take back to the nest and give to its chicks.

PARROTS

Devotees of Tarzan movies may be surprised to learn that this well-known group of brightly coloured and charismatic fruit-eaters is rather poorly represented in Africa, which hosts a mere 19 of the world's 360 Psittacidae species, of which a dozen occur in East Africa. Most African parrots are rather elusive in the wild, often drawing attention to their presence with a series of explosive screeches, then immediately disappearing from view with their characteristic direct flight. Particularly vociferous is that familiar red-tailed caged bird, the African grey parrot (*Psittacus erithacus*), a western forest species whose range extends across Uganda (where it is commonly seen around Entebbe and in Mabira and Kibale forests) into the Kakamega region of western Kenya. More likely to be seen in the region's main game reserves are Meyer's parrot (*Poicephalus meyeri*), which has diagnostic bright yellow shoulders, and its striking eastern counterpart the orange-bellied parrot (*P. rufiventris*).

Fischer's lovebirds nesting in a dead tree. (AVZ)

Lovebirds are sociable and noisy small green parrots with a parakeet-like bill, whose upper mandible almost completely hides the lower one. More likely to be seen on the ground than other parrots, they are normally associated with well-wooded savannah, where they feed mainly on seeds and nest in tree cavities. Most species have a limited distribution: the yellow-collared and Fischer's lovebirds (*Agapornis personatus* and *A. fischeri*) are both more or less endemic to northern Tanzania, with the former most likely to be seen in Tarangire National Park and the latter in the Serengeti ecosystem. Feral populations of both are also well established in the vicinity of Kenya's Lake Naivasha, where they frequently hybridise. The most widespread species north and west of Lake Victoria is the red-faced lovebird (*A. pullarius*).

TURACOS

Endemic to Africa, the turacos or loeries (family Musophagidae) are medium to large fruit-eaters with a characteristic long tail and upright crest. East Africa is home to eight typical turacos of the genus *Tauraco*. These are spectacularly colourful birds of forest or dense woodland, with green backs and dazzling red flight feathers, and often some white markings in the face or crest. Most species are rather similar in appearance and habits, but identification is aided

Great blue turaco is common in the leafy suburbs of Entebbe. (AVZ)

White-bellied go-away bird (AVZ)

by the minimal overlap in their ranges. They tend to be quite furtive birds, disappearing among the branches with a characteristic bounding run, but often draw attention to their presence with far-reaching guttural calls that resonate through the canopy.

Larger than any *Tauraco* species, Ross's turaco (*Musophaga rossae*) is a truly gorgeous apparition, its uniform glossy purple feathering off-set by a bright yellow bill and face mask, and crimson on the crest and wings. It is widespread in the west of the region: good localities include riparian woodland in the Serengeti-Mara ecosystem, Kenya's Saiwa Swamp National Park, and forest-fringed wetlands in Uganda and Rwanda. More magnificent still, the great blue turaco (*Corythaeola cristata*) is a very large (75cm) blue bird with yellow, green and orange markings on its breast, a yellow and red bill, greenish tail panels and an upstanding blue crest. It is restricted to western forests, where flocks of up to ten file clumsily between the trees like psychedelic airborne turkeys, and is particularly characteristic of Rwanda's Nyungwe Forest and Kenya's Kakamega Forest. This species is widespread in Uganda, and is often seen in Entebbe Botanical Garden, close to the international airport.

The comparatively drab go-away birds (*Crinifer* spp.) prefer wooded savannah habitats, where small flocks tend to perch openly, emitting their onomatopoeic nasal call from the treetops. The grey go-away bird (*C. concolor*) is a monotone southern species whose range extends into the southeast of Tanzania. The bare-faced go-away bird (*C. personata*) has a grey back, white chest and black face, and inhabits west-central Tanzania and the Lake Victoria basin. The white-bellied go-away bird (*C. leucogaster*) is grey with a white belly, and inhabits drier parts of Kenya, central Tanzania and northeast Uganda. The eastern grey plantain-eater (*C. zonurus*), the only grey turaco with a shaggy crest and bold yellow bill, is restricted to the northwest, and its infectious chuckling and giggling are characteristic sounds of rural Uganda.

CUCKOOS AND COUCALS

East Africa's 17 cuckoo (Cuculidae) species are all brood parasites, which means they lay their eggs in the nests of other birds – known as hosts – and play no part in incubating their eggs or raising their young. Most species are highly vocal, but their furtive behaviour makes them hard to see. The red-chested cuckoo (*Cuculus solitarius*), which parasitises robin-chats and scrub robins, is among the region's best-known birds thanks to its persistent three-note call – often transcribed as 'it-will-rain'. The call of the pretty green Diederik cuckoo (*Chrysococcyx caprius*), often heard near the nesting weavers that it parasitises, is a clamorous 'dee-dee-dee-diederik' repeated to a hysterical crescendo. The stunning but legendarily elusive emerald cuckoo (*C. cupreus*) is a forest species whose ventriloqual four-note call, a triumphal 'you-can't-find-me', drives birdwatchers to distraction as they scan the leafy canopy in vain.

Less numerous but more visible than the closely related cuckoos, the coucals of the family Centropodidae are large, clumsy birds that frequent rank grassland, marsh, and lake margins. The white-browed coucal (*Centropus superciliosus*) has a clear white eye-stripe and streaked underparts, and is often referred to as the 'rainbird' on account of its attractive bubbling call, which is said to predict rain. Equally distinctive but more closely restricted to marshy habitats, the handsome black coucal (*C. grillii*) is all black except for its rich chestnut wings. The region's other four species are confusingly similar in appearance (chestnut wings and back, black cap, creamy white underparts, red eye) but have more or less mutually exclusive ranges.

White-browed coucal (AVZ)

OWLS

Feared in many African cultures as harbingers of death, the owls (order Strigiformes, families Tytonidae and Strigidae) are the nocturnal equivalent of the raptors, with sharp talons and hooked bills designed for hunting, excellent night vision as a result of their large round eyes, and exceptional hearing. Some 22 species are present, ranging from the thrush-sized scops owls to the gigantic eagle owls. You are, naturally, most likely to see owls at night – especially on night drives – though you may occasionally flush or chance upon some species in daylight. The common barn owl (*Tyto alba*) often announces its presence with a hissing, screeching call and looks spectrally white in nocturnal flight, whereas the equally widespread Verreaux's and spotted eagle owls (*Bubo lacteus* and *B. africanus*) are much larger and browner, and have softer grunting and hooting calls. Of the smaller owls, the rapacious pearl-spotted owlet (*Glaucidium perlatum*) is a partially diurnal hunter that is frequently mobbed by other small birds, while the African scops owl (*Otus senegalensis*) has a remarkable ability to disguise itself by flattening its bark-coloured body against a tree trunk.

Four East African owls are regional endemics: the Usambara eagle owl (*Bubo (poensis) vosseleri*) of the Usambara and Uluguru mountains; the Pemba scops owl (*Otus pembaensis*) of Pemba Island;

the Sokoke scops-owl (*Otus ireneae*) of western Kenya's Sokoke Forest; and the Albertine owlet (*Glaucidium albertinum*) of the eastern Congo and Rwanda's Nyungwe Forest. None of these, however, is as eagerly sought by birdwatchers as the localised Pel's fishing owl (*Scotopelia peli*), an impressive rufous-orange, eagle-sized resident of dense riverine woodland, which feeds almost exclusively on fish. The Itombwe owl (*Tyto prigoginei*) was described by science in 1951, based on a single skin collected in the Itombwe Massif (DRC), and has been recorded only once since, when it was mist netted in the same area in 1996. Also known as Congo bay owl, this mega-rarity is assumed, by truly optimistic birdwatchers, to have been the source of an unidentified owl call recorded in Nyungwe Forest in 1990.

A daylight glimpse of the pearl-spotted owlet, one of the region's smaller and more diurnal owl species. (AVZ)

Spotted eagle-owl (AVZ)

NIGHTJARS

Nightjars (Caprimulgidae) are dove-sized nocturnal birds that feed exclusively on flying insects, which they hawk in mid-air using their prodigious gape. Most of East Africa's 17 species look very similar, and they are most often observed on roads after dark, when their cryptic coloration is distorted by the artificial cast of the spotlight or headlight. The most reliable clue to identification lies in their distinctive calls, none more so than that of the fiery-necked nightjar (*Caprimulgus pectoralis*), whose quavering seven-syllable warble (often rendered as 'good-lord-deliver-us') might be repeated 100 times in short succession.

Two nightjar species become unmistakable when the male enters its astonishing breeding plumage between August and March. The more widespread is the pennant-winged nightjar (*C. vexillarius*), which is often seen flying above water at dusk, trailing elongated second primary feathers that form a pair

of pennants twice as long as its body. The standard-winged nightjar (*C. longipennis*) is confined in East Africa to northwest Uganda, where it is most often seen in Murchison Falls. The male's extended second primaries terminate in leaf-like 'flags', which – when the bird takes to the air – create the impression that two small birds are chasing a larger one!

SWIFTS

Superficially similar to the unrelated swallows, the swifts (family Apodidae) are the most aerially specialised of birds, capable of roosting on vertical surfaces only, and generally feeding and mating while in flight – indeed, some species might go months without touching terra firma, snatching all the sleep they need while on the wing. Swifts tend to circle in the vicinity of cliffs, bridges and tall buildings, where large mixed flocks might include several species – often intermingled with swallows, from which they can be distinguished by their

Above Mozambique nightjar (*Caprimulgus fossii*) (AVZ)
Below European swift (*Apus apus*) (AVZ)

narrower, crescent-shaped wings. Most swifts are dull grey-brown in colour, but some, such as the oversized alpine swift (*Tachymarptis melba*) and white-rumped swift (*Apus caffer*) have distinctive and diagnostic white markings. Spinetails, named for the row of tiny spikes on the tips of their squared tails, are less strictly aerial than other swifts and are often seen fluttering bat-like in the vicinity of baobab trees.

MOUSEBIRDS

The mousebirds (family Coliidae) are endemic to Africa. These long-tailed and prominently crested fruit-eaters shuffle nimbly among branches in the manner of their namesakes – although their name might equally refer to their mousy coloration. Generally seen in small tight flocks, the rather scruffy speckled mousebird (*Colius striatus*) is the most common of four species found in East Africa.

TROGONS

Trogonidae, a predominantly South American and Asian family of colourful insectivores, is represented in East Africa by two eagerly sought species. The Narina trogon (*Apaloderma narina*) has a lustrous green back and wings, rich crimson breast, yellow bill and plain tail, while the more localised bar-tailed trogon (*A. vittatum*), though similar in overall appearance, has a much darker head, a blue chest band above its red breast, and a densely barred underside to the tail. Both birds tend to perch motionless on a horizontal branch, often high in the canopy, occasionally betraying themselves with their distinctive calls. The Narina trogon is an elusive resident of riparian woodland and forest interiors throughout the region, while the bar-tailed trogon is associated primarily with highland forest – it is quite common in Uganda's Bwindi Impenetrable National Park and is likely to be seen on any forest walk led by an ornithologically astute local guide, especially one who is able to imitate its call.

KINGFISHERS

East Africa's 15 kingfisher species (family Alcedinidae), are a characterful and varied bunch, but all are easily recognisable as kingfishers thanks to their stocky build, upright stance and long dagger-like bill. They range in size from the 45cm giant kingfisher (*Megaceryle maxima*), a pied, chestnut-breasted bird associated with tree-lined rivers and lakes, to the tiny African dwarf kingfisher (*Ispidina lecontei*), an orange, blue and pink bird of Ugandan forest interiors. The most visible species of aquatic habitats is the unmistakable pied kingfisher (*Ceryle rudis*), a starling-sized back-and-white bird, which hunts fish by hovering above the open water then diving down to grab its prey. Also very common, the gemlike malachite kingfisher (*Corythornis cristata*) resembles the familiar European species, but is even smaller and has

Above Speckled mousebird (AVZ)
Below The giant kingfisher is the world's largest kngfisher. (AVZ)

167

a bright red bill. It is often observed perched motionless on waterside reeds and shrubs, typically less than a metre above the surface.

Not all kingfishers eat fish. The most characteristic African kingfishers are the seven species of the genus *Halcyon*, most of which have brilliant blue wings, pale underparts and strong black and/or red bills. Aside from the coastal mangrove kingfisher (*H. senegaloides*), these insect-eaters frequent wooded habitats – often some distance from water. Widespread species are the striking grey-headed kingfisher (*H. leucocephala*), with its bright chestnut belly; the dumpier striped kingfisher (*H. chelicuti*), with its streaked underparts; and the pale blue woodland kingfisher (*H. senegalensis*), an intra-African migrant whose explosive rattling call resounds through the wooded savannahs of Tanzania and Uganda between September and April.

BEE-EATERS

The dashing and dazzlingly colourful bee-eaters (family Meropidae) are well represented in East Africa, with 17 of the world's 25 species being present. All bee-eaters have an upright stance and distinctive sleek profile, with long wings and tail, and long slightly decurved bills. Most are sociable hole-nesters associated with wooded savannah and riverine habitats, where they perch openly on bare branches, sporadically darting off to catch an insect on the wing before returning to the same (or a nearby) perch. All bee-eaters are capable of eating stinging insects, which they disarm by banging or scraping against a branch, but they will also often snap up winged termites and other harmless flying insects.

East Africa's bee-eaters are all placed in the genus *Merops*, and all but five are predominantly

The malachite kingfisher (*top*, AVZ) is usually seen perching low on waterside vegetation, whereas the grey-headed kingfisher (*centre*, AVZ) is a woodland species that feeds mostly on insects. The striking pied kingfisher (*bottom*, AVZ) is often seen hovering over the shallows of large lakes and other aquatic habitats.

The exquisite white-fronted bee-eater (*main picture,* AVZ) and gorgeous carmine bee-eater (*inset,* AVZ) are among the region's most striking birds.

green with a black bill and narrow black face mask. Nevertheless, each can be identified by its combination of size, tail shape, throat colour and presence or absence of red or blue markings. The ubiquitous little bee-eater (*Merops pusillus*) might be seen anywhere in the region, usually in pairs, which generally stick to woodland habitats along lake margins and watercourses. The Nile below Murchison Falls is a good spot for the stunning red-throated bee-eater (*M. bulocki*), while the equally handsome white-fronted bee-eater (*M. bullockoides*) is common in southern Tanzania and the Kenyan Rift.

Atypical members of the genus include the migrant European bee-eater (*M. apiaster*), which sports a stunning combination of golden-brown back, yellow throat and blue breast; the localised forest-dwelling black- and blue-headed bee-eaters (*M. gularis* and *M. muelleri*), which are both predominantly black or blue with scarlet throats; and the stunning northern and southern carmine bee-eaters (*M. nubicus* and *M. nubicoides*), which are bright red with blue heads and bellies. The last of these will sometimes hitch a lift on the back of large ground birds such as the kori bustard in order to snap up insects disturbed around its feet.

169

The lilac-breasted roller habitually chooses conspicuous perches. (AVZ)

ROLLERS

Named for their agile aerial displays, the dazzling rollers (family Coraciidae) are robust, pigeon-sized insectivores that perch openly in savannah country, occasionally dropping to the ground to swoop on an item of prey. The lilac-breasted roller (*Coracias caudatus*) is the acknowledged star of the family, and one of the most popular and instantly recognisable safari birds, rather jay-like with its lilac chest, sky-blue underparts and golden-brown back. Four other similar looking species occur in East Africa's savannahs, including the European roller (*C. garrulus*), which is locally common during the northern winter. The broad-billed and blue-throated rollers (*Eurystomus glaucurus* and *E. gularis*) are smaller forest birds, with dark purple, blue and brown plumage and conspicuous yellow bills.

HOOPOES, WOOD-HOOPOES AND SCIMITAR-BILLS

Upupiformes is a recently created order that comprises three families, two of which are endemic to Africa. The best-known species is the African hoopoe (*Upupa africana*), a distinctive pinkish-orange bird with a long curved bill, striking black-tipped crest and floppy black-and-white wings. It is common in park-like savannah, as well as in suburban hotel gardens – where it is often seen in a nodding walk probing the lawn for invertebrates – but is replaced by the very similar Eurasian hoopoe (*Upupa epops*) in the north of the region.

Very different in appearance, wood-hoopoes are gregarious, glossy-black woodland birds with long tails and decurved red or black bills. The most widespread species is the green wood-hoopoe (*Phoeniculus purpureus*),

African hoopoe (AVZ)

170

which travels from tree to tree in small, cackling flocks, clambering acrobatically over the trunk and branches as they search for grubs beneath the bark. Similar in general appearance, but smaller, quieter and less gregarious, the three scimitar-bill species (genus *Rhinopomastus*) are common within their respective ranges, which collectively cover most corners of East Africa.

HORNBILLS

Bucerotidae is a conspicuous family of characterful and often rather comical medium-to-large birds that inhabit practically every habitat from desert to jungle, and can easily be recognised by their trademark heavy decurved bills. Most of the region's 23 species nest in holes in tree trunks, with the female sealing herself in by plastering closed the entrance for the full incubation period, during which time she undergoes a full moult, while the male feeds her through a small slit until the eggs have hatched and the chicks are growing larger, when she breaks out of her self-created prison.

East Africa's most conspicuous hornbill species are mostly ground-feeding omnivores, and they include such firm safari favourites as the eastern yellow-billed hornbill (*Tockus flavirostris*), the northern red-billed hornbill (*T. erythrorhynchus*), Von der Decken's hornbill (*T. deckeni*) and the African grey hornbill (*T. nasutus*). Slightly larger and darker, the crowned hornbill (*Lophoceros alboterminatus*) is a widespread woodland bird with a casqued red bill, while the Congo pied hornbill (*L. fasciatus*) is a yellow-billed forest bird of south-central Uganda. The recently described Tanzanian red-billed hornbill (*T. ruahae*) is endemic to Ruaha National Park and environs.

The fruit-eating forest hornbills of the genus *Bycanistes* are much bulkier, and predominantly black-and-white in colour, with enormous casqued bills. Among the most conspicuous residents of the East African forest, these outsized birds generally occur in small flocks that reveal their presence with raucous nasal calls and heavy wingbeats. The silvery-cheeked hornbill (*B. brevis*) of the central Kenyan and eastern Tanzania highlands is replaced by the black-and-white casqued hornbill (*B. subcylindricus*) in western Kenya and Uganda, and by the trumpeter hornbill (*B. bucinator*) in some parts of Tanzania and southeast Kenya. Several other forest hornbills have very restricted ranges in East Africa, with four species being confined to Semuliki National Park in western Uganda.

The Tanzanian red-billed hornbill, endemic to central Tanzania, has only recently been described. (AVZ)

A massive southern ground hornbill takes to the air in Serengeti National Park, Tanzania. (AVZ)

Ground hornbills are decidedly improbable turkey lookalikes, whose general black coloration is off-set by striking white primaries (best seen in flight), large casqued bills, conspicuous throat and eye wattles, and long fluttering eyelashes. Typically seen strutting through the savannah in parties of up to five birds, these bulky predators are often assumed to be flightless, but they are surprisingly strong fliers, and both roost and nest in trees – although, unlike other hornbills, the female does not seal herself into the nest chamber. The southern ground hornbill (*Bucorvus leadbeateri*), a common resident of game reserves in Tanzania and southern Kenya, has all red wattles, whereas the Abyssinian ground hornbill (*B. abyssinicus*), of northern and western Kenya and Uganda, has blue eye wattles and a blue or blue-and-red neck wattle. Both are among the earliest voices of the pre-dawn chorus, making a deep, rhythmic booming call that can sound almost like a distant lion.

WOODPECKERS

East Africa has 20 species of woodpecker (Picidae). Most are almost exclusively arboreal, using their sturdy bill to chisel out insects from beneath the bark of a tree while clinging upright to its trunk or branches with their zygodactyl feet (two toes facing forward; two facing back) and stiffened tail. They tend to be rather inconspicuous when feeding, but some species announce their presence with a strident call or – during courtship – by drumming loudly on a dead branch.

Most of the region's woodpeckers conform to a similar plumage pattern, with dark green upperparts, paler

Grey woodpecker (AVZ)

underparts and some red on the crown. Individual species can be identified by the precise arrangement of the (often rather complex) head markings and belly patterning. Among the most common are the cardinal woodpecker (*Dendropicos fuscescans*), a sparrow-sized woodland bird with heavily streaked underparts, and the bearded woodpecker (*Chloropicus namaquus*), a starling-sized bird with a darkly barred belly. Rather atypical insofar as they have grey or green unmarked bellies, the grey and olive woodpeckers (*D. goertae* and *D. griseocephalus*) are associated with moist savannah in the north of the region and montane forests in the south respectively.

BARBETS AND TINKERBIRDS

Closely related to woodpeckers but rather different in appearance, the barbets (family Capitonidae) are stocky, small to medium-sized birds, represented in East Africa by 33 species that collectively span every habitat from desert to rainforest. The smallest are the finch-sized tinkerbirds of the genus *Pogoniulus*, a group of nine delicately marked yellow, black and white species associated with forest and woodland, where they tend to perch high in the canopy and repeat tirelessly their piping one-note call.

The most common barbets of drier savannah habitats are the ground-dwelling *Trachyphonus* species. These include the red-and-yellow barbet (*T. erythrocephalus*), a strikingly coloured bird that is often seen performing its dramatically daft clockwork-like duet on termite mounds in Tarangire or Samburu-Buffalo Springs. Its southern equivalent is the crested barbet (*T. vaillantii*), whose high-pitched trilling call is characteristic of Ruaha National Park.

Red-and-yellow barbets sitting on a termite mound, from which they feed. (AVZ)

Also very spectacular are the region's eight species of the closely allied genera *Lybius* and *Pogonornis*. These black, white and red birds include the stunning double-toothed barbet (*P. bidentatus*) of Uganda and vociferous black-collared barbet (*L. torquatus*) of Tanzania and Kenya. The red-faced barbet (*L. rubrifacies*) is a regional endemic restricted to moist savannahs west of Lake Victoria, where it may be observed in Rwanda's Akagera National Park. The yellow-billed barbet (*Trachyphonus purpuratus*) is a stunning long-tailed purple, black and yellow forest species, and most likely to be seen in the vicinity of Kibale National Park in Uganda.

The greater honeyguide often spends hours repeating its explosive two-note call from an open perch. (SS)

HONEYGUIDES

Honeyguides are inconspicuous and nondescript relatives of the barbets and woodpeckers, represented in East Africa by 14 species, of which the most striking is the greater honeyguide, which looks a bit like a large sparrow with a heavy pink bill. A widespread and common resident of riparian and other woodland habitats, the greater honeyguide is the only member of the family that has been confirmed to perform the eponymous trick of guiding people (and, possibly, honey badgers) to beehives, where it will feed on any discarded scraps of honeycomb. All honeyguides are brood parasites, with most species favouring the hole-nesters such as the related barbets and woodpeckers as hosts, though some smaller forest species parasitise warblers and white-eyes.

PASSERINES

Passeriformes is arguably the most successful of vertebrate orders, with roughly 5,400 species accounting for more than half of the global avifauna. Passerines are also known as perching birds, and technically speaking they are distinguished from other birds by their unique arrangement of one backward-pointing and three forward-facing toes that are all similar-sized and meet the foot at the same level, which allows them to grip tightly around a perch. Most passerines are smaller than a dove, the main exception being the crow family, which includes several larger species. Many are also impressive songsters.

Such is the sheer number of different passerines in East Africa that the descriptions below operate mainly at family and genus level. For practical reasons, the various passerine families are arranged in line with the taxonomic conventions followed by traditional field guides. It is worth noting, however, that DNA hybridisation techniques pioneered in the 1990s suggest that many such conventions possess little genetic validity. Doubtless, these advances in our understanding of avian taxonomy will eventually be more widely detailed in field guides and the like, but for the time being they are too controversial and specialised to warrant more than passing mention here.

LARKS

The larks (family Alaudidae) are small, solitary and easily overlooked ground birds that frequent short grassland and semi-arid habitats, and tend to be difficult to tell apart in the field. The most distinctive and widespread of 20-odd East African species is the rufous-naped lark (*Mirafra africana*), which spends long periods repeating its drawn-out three-note song from treetops or other prominent perches. Somewhat more distinctive are the region's three *Eremopterix* (sparrow-lark) species, all of which are relatively gregarious and have bold black-and-brown sparrow-like markings.

SWALLOWS AND MARTINS

The swallow family (Hirundinidae) is represented in East Africa by at least 23 species of medium-small, highly aerial passerines that somewhat resemble swifts, but are generally more colourful. The true swallows of the genus *Hirundo* – which include the European barn swallow (*H. rustica*), an abundant Palaearctic migrant from August to April – are attractive birds, predominantly blue in appearance but with white, red or streaked underparts. Many have colourful throats or rumps and sport elongated tail streamers. Mixed swallow flocks often swoop and dart above water, sometimes in the company of martins, which have a similar shape but are generally more brown than blue. Some species, for instance the brightly coloured mosque swallow (*Cecropsis senegalensis*), are most often seen in pairs in bush country, where they often roost in exposed treetops. The saw-wings (genus *Psalidoprocne*) are black or pied swallows associated with forest habitats.

WAGTAILS, PIPITS AND LONGCLAWS

Motacillidae is a family of predominantly terrestrial birds whose slender appearance is accentuated by a narrow bill and, in wagtails, a long tail. Wagtails (genus *Motacilla*) are boldly marked birds usually observed strutting purposefully along riverbanks and lakeshores, tail incessantly bobbing up and down. Most common is the resident African pied wagtail (*M. aguimp*), though this is outnumbered during the northern winter by the migrant yellow wagtail (*M. flava*), of which seven distinct races are recognised.

Top Rufous-naped lark (AVZ)
Centre Lesser striped swallow (AVZ)
Bottom Yellow-throated longclaw
(*Macronyx croceus*) (AVZ)

East Africa also harbours a dozen pipits of the genus *Anthus*, most of which are predominantly streaky brown in colour and difficult to identify on a specific level, as well as the golden pipit (*Tmetothylacus tenellus*), a boldly marked dry-country bird with a bright yellow head and bold black bib. Endemic to Africa, the longclaws of the genus *Macronyx* are pipit-like grassland birds with striking yellow, orange or pink throats, black bibs, and in some cases yellow bellies.

BULBULS AND ALLIES

One of the first birds to be encountered by most visitors to East Africa, the versatile common bulbul (*Pycnonotus barbatus*), also known as the black-eyed or yellow-vented bulbul, is a cheerful and confiding medium-small garden bird with an upright stance, a black crested head and a trademark yellow vent. The related greenbuls and brownbuls, also placed in the family Pycnonotidae, are represented by at least 35 East African species. These are forest birds with muted green, grey, brown and/or yellow feathering, and pose much more of an identification problem – especially as many forests host a dozen or more species, whose nuances of coloration are often lost in the gloomy light. Experienced birdwatchers find it easiest to distinguish most by their voices.

BABBLERS AND CHATTERERS

The babblers and chatterers (family Leiothrichidae) are robust, thrush-sized birds that habitually travel through the undergrowth in noisy parties of 5–10 individuals. Ten species are present in East Africa, of which the arrow-marked babbler (*Turdoides jardineii*) and black-lored babbler (*T. sharpei*) are most commonly seen on safari, though the brown babbler (*T. plebejus*) is commoner in parts of Uganda. Other notable species are Hinde's babbler (*T. hindei*), a vulnerable endemic of the lush Kenyan highlands north of Nairobi, and the rufous chatterer (*Argya rubiginosa*), a slender orange-brown dry-country bird of northern Tanzania and Kenya.

Below left Common bulbul (AVZ)
Below right Rufous chatterer (AVZ)

THRUSHES

Turdidae is a diverse group of colourful medium-to-small insectivores, of which the most familiar will be the *Turdus* thrushes, which – like their European counterparts – are common garden birds that hop openly over lawns in search of worms. The six resident species in East Africa all have brownish backs, greyish chests, pale orange bellies and orange or yellow bills. Two very localised endemics, the Taita and Usambara thrushes (*T. helleri* and *T. roehli*), are confined to the mountain ranges for which they are named.

The robin-chats typically have bright orange underparts, brownish or blueish upperparts, and in most cases a bold white eye stripe or cap. Denizens of forest, scrub and suburban gardens, robin-chats can become very confiding around people, and tend to be exuberant songsters and skilled mimics. Commoner species include the pastel-shaded Cape robin-chat (*Cossypha caffra*), the larger and more boldly marked white-browed robin-chat (*C. heuglinii*) and the very pretty red-capped robin-chat (*C. natalensis*). Robin-chat diversity is highest in Uganda, where a quartet of western forest species occurs at the easternmost part of their range.

The white-starred robin (*Pogonocichla stellata*), a widespread but shy resident of East Africa's montane forests, is a striking blue-headed and yellow-chested skulker with a white spot on its chest. Also primarily associated with forest interiors are the eight akalat species (genus *Sheppardia*) and four species of alethe, most of which are rather localised and nondescript – one exception being the red-throated alethe (*Chamaetylas poliophrys*), an Albertine Rift endemic commonly encountered in Nyungwe Forest.

Above White-browed robin-chat (AVZ)
Below Sooty chat (*Myrmecocichla nigra*) (AVZ)

Old World flycatchers also include several genera associated with savannah, grassland, rocky slopes and dry-country habitats. The wheatears (genus *Oenanthe*) are small, slender upright birds of open country, represented in East Africa by five Palaearctic migrant and three resident species, including the endemic Schalow's wheatear (*O. (lugubris) schalowi*). Similar in shape and stance, two genera of chats are widely represented in East Africa: the dark brown, wing-flicking *Cercomela*, and the more boldly marked *Myrmecocichla*. The mocking cliff chat (*Thamnolaea cinnamomeiventris*) is a striking relative of the latter, and is often encountered flaunting its black, white and red plumage around rocky koppies.

The African paradise flycatcher feeds on winged insects ranging in size from flies to butterflies. (AVZ)

A common resident of open grassland, where it tends to perch on top of bushes and low trees, the African stonechat (*Saxicola torquatus*) is a neat pied chat with a bold chestnut breast, while the related whinchat (*S. rubetra*) is a Palaearctic migrant that can be very common in the moist grasslands of western Uganda between October and April. More terrestrial than chats, and less upright in stance, the scrub-robins and palm-thrushes of the genera *Cercotrichas* and *Cichladusa* are represented by six and two species respectively, of which the white-browed scrub-robin (*Cercotrichas leucophrys*) and spotted palm-thrush (*Cichladusa guttata*) are most likely to be seen on safari.

FLYCATCHERS
Monarch flycatchers

Unrelated to the Old World flycatchers (Muscicapidae), monarch flycatchers (Monarchidae) are generally active and colourful crested birds which combine hawking insects with gleaning them from leaves. The best-known monarch is the hyperactive African paradise flycatcher (*Terpsiphone viridis*), which occupies most habitats other than true desert, including suburban gardens. This spectacular species comes in white, rufous and intermediate morphs, with the male's splendid tail in all cases growing to be three times as long as its body.

Old World flycatchers

The Old World flycatchers (Muscicapidae) is a diverse family of insect-eaters which include many very familiar species. Typically, flycatchers are dull and rather inconspicuous birds that hawk insects from static perches, but several other bird groups formerly classified with the Turdidae – among them robin-chats, alethes, wheatears and chats – are now included in the Muscicapidae.

Left African grey flycatcher (*Bradornis microrhynchus*) (AVZ)
Right Grey-backed camaroptera (*Camaroptera brachyura*), a member of the warbler family (AVZ)

WARBLERS AND OTHER SMALL INSECT EATERS

The warblers and allies (Sylviidae) are traditionally thought to comprise some 550 species worldwide, though they are one of the bird families most radically affected by recent advances in avian taxonomy. Some authorities argue that genera such as *Elminia* 'flycatchers' and *Turdoides* babblers should be included in Sylviidae, while other genera such as *Cisticola* and *Prinia* should be reassigned to a separate family, Cisticolidae. Until such time as these taxonomic controversies are settled, around 135 of East Africa's bird species are conventionally assigned to Sylviidae, most of them rather small, nondescript and inconspicuous LBJs ('little brown jobs') with narrow pointed bills typical of insect eaters.

Perhaps the most confusing of all African bird genera, the cisticolas (genus *Cisticola*) comprise around 35 species of small brown grassland birds, most of which can be identified by call and behaviour only. By contrast, the apalises (genus *Apalis*) consist of 18 species of long-tailed, boldly marked leaf gleaners, while the eight crombec species (genus *Sylvietta*) are remarkable for their almost tailless appearance. The family is littered with striking one-offs, too: the splendid black-faced rufous warbler (*Bathmocercus rufus*) is a forest-fringe species often seen in Bwindi Impenetrable National Park; the endearing grey-capped warbler (*Eminia lepida*) is a widespread resident of waterside thickets; and the localised red-winged grey warbler (*Drymocichla incana*) is most likely to be seen in riverine woodland in Murchison Falls.

The superficially warbler-like white-eyes (family Zosteropidae) are small green-yellow birds with bold white eye-rings. They are usually seen moving through the foliage in parties of five to ten individuals and are common in suburban gardens. Represented in East Africa by 11 species, the tits (family Paridae) are active little savannah birds that spend much of their time working their way along tree trunks gleaning insects, often hanging upside down to do so. The true tits of the genus *Melaniparus* are canary-sized black-and-white birds that often join mixed bird parties, while the penduline tits (family Remizidae) are tiny, nondescript birds whose elaborate hanging nests possess a false entrance and dummy nest chamber to fool would-be nest raiders.

Long-tailed Fiscal (*Lanius cabanisi*) (AVZ)

SHRIKES

Sometimes referred as butcher birds, in reference to their custom of impaling prey on a thorn bush or barbed wire fence, the true shrikes (genus *Lanius*) are conspicuous and boldly marked perching birds with a characteristic upright stance, a heavy hooked bill and a longish tail. The fiscal shrikes comprise six resident black, white and in some cases grey species, of which the common fiscal (*L. collaris*) is most widespread and typically associated with medium to high altitudes, while the sociable grey-backed fiscal (*L. excubitoroides*) prefers moist savannah habitats. During the northern winter, shrike numbers are boosted by several migrant species from Europe, such as the red-backed shrike (*Lanius collurio*). The most striking of the true shrikes is the magpie shrike (*L. melanoleucus*), a sociable black savannah species with white wing markings and a long tail that flops around loosely in the breeze.

Drongos and cuckoo-shrikes

One of the most characteristic and characterful passerines of savannah and woodland habitats, the fork-tailed drongo (*Discrurus adsimilis*) perches openly like a large shrike, but differs from any shrike in being all black with a deeply forked tail, and has a wide array of nasal calls. It is a highly active and aggressive bird, and will often mob raptors and other animals it considers to be a threat. Although two other drongo species are present in the region, they are confined to forest habitats, so the fork-tailed drongo is most likely to be confused with one of the black flycatchers – which are more lightweight birds with an unforked tail. Also confusable with drongos, cuckoo-shrikes are inconspicuous woodland and forest birds that show an unusually high level of sexual dimorphism, with four of the six East African species being predominantly black (male) or yellow (female), and two being all grey.

Fork-tailed drongo (AVZ)

Bush-shrikes

Endemic to mainland Africa, the bush-shrikes (Malaconotidae) are a family of medium-small shrike-like birds with strong hooked bills, striking colours and somewhat furtive habits – many species would easily be overlooked were it not for their distinctive and far-carrying calls. The skulking bush-shrikes are brilliantly coloured birds of savannah woodland and forest, typically with green backs, yellow underparts, some grey on the head, and in several cases a black mask and/or bib and a red or orange throat. The acacia-loving sulphur-breasted bush-shrike (*Telophorus sulfureopectus*) often joins mixed feeding parties of birds, while the sturdier grey-headed bush-shrike (*Malaconotus blanchoti*) is sometimes called the ghost-bird in reference to its frustrating habit of sitting motionless in the high canopy, repeating its drawn-out hollow whistle ad infinitum.

Slate-coloured boubou (AVZ)

181

The best-known members of this family are the boubous of genus *Laniarius*, whose tightly synchronised duets typically involve the male singing a short series of loud antiphonal notes and the female replying with a harsh clicking or chirring sound. The black-and-white tropical boubou (*Laniarius major*), often known as the bellbird in reference to its ringing call, is common in moist parts of the region, where it is often seen in gardens, while the all-black slate-coloured boubou (*L. funebris*) is associated with drier bush habitats. Common in much of Uganda, the black-headed gonolek (*L. erythrogaster*) is a striking boubou with bright crimson underparts, as is the more elusive swamp-dwelling papyrus gonolek (*L. mufumbiri*), which also has a white wing bar and yellow cap.

Also present in the region are four species of tchagra (genus *Tchagra*), which have chestnut backs, pale bellies and white eye-stripes. They are typically seen skulking in thick bush at knee level or lower, but are conspicuous by their melodious songs and displays. The less secretive puffbacks (genus *Dryoscopus*) are small, pied, red-eyed bush-shrikes named for the male's habit of puffing up his white back feathers like a snowball when displaying. The widespread brubru (*Nilaus afer*) is a puffback-like bird whose slurring call sounds a bit like an old-style telephone.

Malaconotidae is now placed in a superfamily that also includes the taxonomically controversial Prionopidae and Platysteiridae. Endemic to Africa, the family Prionopidae, now thought to be closely related to the vangas of Madagascar, consists of five species of helmet-shrike, which are medium-small black or pied birds with conspicuous red or yellow eye-rings, typically observed flying between trees in small

parties. Helmet-shrikes are closely allied to the African shrike-flycatcher (*Megabyas flammulatus*) and black-and-white shrike-flycatcher (*Bias musicus*), a pair of odd glossy-black forest birds with wonderful display flights, as well as to the vangas of Madagascar. Platysteiridae is represented in East Africa by nine species of *Batis*. These active little woodland birds, usually seen in pairs, display sexual dimorphism in their black, white, grey and chestnut markings, and usually have yellow or red eyes. The family also includes five species of wattle-eye (similar, but with prominent red or blue eye wattles).

CROWS AND ORIOLES

Crows and ravens (family Corvidae) are familiar birds the world over, known for their predominantly black plumage, powerful bills and scavenging habits. In East Africa they are predominantly birds of dry country, savannah and montane regions, where they spend much of their time on the wing. Ravens are the largest of the group, and East Africa's 56cm white-necked raven (*Corvus albicollis*) is the region's largest passerine, with a hefty bill and diagnostic white nape of the

White-necked raven (AVZ)

neck. Most other East African species, including the black-and-white pied crow (*C. albus*) and fan-tailed raven (*C. rhipidurus*), are also very bulky. Crows are adaptable feeders and regarded as being among the most intelligent of birds, using tools to break open eggs and tortoise shells, and often becoming quite aggressive at campsites.

Loosely affiliated to crows, though much smaller, orioles are brightly coloured, starling-sized birds, typically with a yellow body off-set by a black head or mask, and a red bill. The most widespread species is the African black-headed oriole (*Oriolus larvatus*), a common woodland bird with an attractive liquid call. More exciting for birdwatchers is the green-headed oriole (*O. chlorocephalus*), a montane forest species most likely to be seen in the Eastern Arc Mountains of Tanzania.

African black-headed oriole (AVZ)

STARLINGS

Represented in East Africa by 31 species spilt across nine genera, the starlings (Sturnidae) are a diverse lot in terms of morphology, coloration and habits. Thankfully most species are instantly recognisable as starlings due to their distinctive semi-upright stance, terrestrial habits and strong pointed bills. The most characteristic East African genus is undoubtedly *Lamprotornis*, which includes numerous species of mostly savannah-dwelling 'glossy starlings', including that colourful safari favourite, the aptly named superb starling (*L. superbus*). Also notable are the ashy starling (*L. unicolor*), a Tanzania endemic, and the utterly breathtaking golden-breasted starling (*L. regius*), a dry-country specialist found at Tsavo and Samburu in Kenya.

The superb starling is highly conspicuous in northern Tanzania and southern Kenya. (AVZ)

The golden-breasted starling frequents dry country in Kenya. (AVZ)

The male red-chested sunbird is far more colourful than the rather drab female. (AVZ)

Also noteworthy are the five species of red-winged starling (genus *Onychognathus*), which are associated with cliff or forests; the widespread but unobtrusive violet-backed (or plum-coloured) starling (*Cinnyricinclus leucogaster*) and the ubiquitous and highly gregarious wattled starling (*Creatophora cinerea*).

Closely related to the starlings, the red-billed and yellow-billed oxpeckers of the genus *Buphagus* are nondescript birds that spend much of their time hitching around on the backs off buffalo, antelope, giraffe and domestic cattle, though these days they are quite rare outside of game reserves. It was long thought that oxpeckers fed almost exclusively on ticks and other bloodsucking parasites, and that their relationship with buffaloes and other large mammals was symbiotic, with the host animals benefiting from what is effectively a free parasite-removal service. Recent research indicates that oxpeckers actually play an insignificant role in controlling ticks, but continue to work wounds made by ticks they removed to extract further blood for themselves. In other words, their relationship with large mammals is possibly more parasitic than symbiotic.

SUNBIRDS

Sunbirds (family Nectariniidae), with their iridescent plumage and fondness for flowers, appear to be the Old World equivalent of America's hummingbirds – though they are actually very distant in evolutionary terms. These small, restless passerines all have distinctive long (and in most cases strongly decurved) bills with which to sip nectar from flowers. Almost half of the world's 120 species have been recorded in East Africa, and most show a high degree of sexual dimorphism, with the jewel-like males typically being larger and far more brightly coloured than the females. Beautiful as the males are, positive identification of all but the most singular sunbird species can be a tricky process, especially as the birds seldom sit still for long, and almost half the region's species conform to what might be termed a typical male sunbird plumage, which consists of a green back and head, and a white, cream, black or even yellow belly, separated by one or more coloured bands.

SEEDEATERS

The name seedeaters is applied to a variety of loosely allied bird families (most familiarly sparrows and finches) that feed on small, hard vegetarian fare such as seeds, grains, pips, pits and nuts. Seedeaters can typically be recognised by their heavy conical bills, which enable then to crush hard kernel-like exteriors. Almost all seedeaters are small and many species are rather nondescript, but others – notably the waxbills, canaries, and male widows, weavers and bishops in breeding plumage – are highly colourful and conspicuous.

Female black-necked weaver (*Ploceus nigricollis*) at a partially constructed nest. (AVZ)

Weavers

The weavers of the family Ploceidae are a quintessential part of the African landscape. Common and highly visible in virtually every habitat from rainforest to desert, they are best known for their nest-building skills – to which they owe their common name. More than 75 species are represented in East Africa. This list includes 40 members of the genus *Ploceus* (true weavers), which are among the most characteristic of all African bird genera.

Slightly larger than sparrows, *Ploceus* weavers typically display strong sexual dimorphism, with males being far more colourful than females. Most males conform to the 'masked weaver' colour pattern: predominantly yellow, with streaky back and wings, and a distinct black facial mask, often bordered orange. A dozen East African species fit this prototype more or less absolutely, and a similar number approximate it rather less closely, by having a chestnut-brown mask, or a full black head, or a black back, or being more chestnut than yellow on the belly.

Like the masked weavers, the golden weavers are brilliant yellow and/or light orange with some light streaking on the back, but they lack a mask or other strong distinguishing features. *Ploceus* weavers, which frequent forest habitats, tend to have very striking markings, and the female is often as boldly marked as the male. The most atypical among these is Vieillot's black weaver (*Ploceus nigerrimus*), the male of which is totally black except for its eyes, while the black-billed weaver (*P. melanogaster*) inverts the 'masked weaver' prototype by being all black with a yellow face-mask.

Bishops and widows

Also belonging to the weaver family in East Africa are 14 species of the genus *Euplectes*, striking black, red and yellow birds that include the larger widows of marsh and moist grassland, and the smaller bishops of reed-lined wetlands. The star of the genus, often to be seen in low laboured display flight over rank grassland, is undoubtedly the long-tailed widow (*Euplectes progne*), a medium-sized black weaver with red

Top Speckle-fronted weaver (*Sporopipes frontalis*) (AVZ)

Below White-headed buffalo-weaver (*Dinemellia dinemelli*) (AVZ)

shoulder markings and, during the breeding season, a spectacular droopy tail that gives the bird a total length of 75–80cm. Also very striking are the region's four red bishop species, which breed communally in reedbeds, and the quirky golden bishop (*E. capensis*), which puffs itself up to look like an oversized bumblebee buzzing above the marshes where it breeds.

A male golden palm weaver, *Ploceus bojeri*, perches next to a freshly constructed nest. (AVZ)

MASTERS OF DESIGN

Weavers are justly famed for the elaborate ball-shaped nests woven by the dextrous males. Once a nest site is chosen, usually at the end of a thin branch, it is stripped of leaves to allow nowhere for snakes to hide. The weaver then flies back and forth, carrying the building material blade by blade in his heavy beak, first hanging a skeletal ring-like nest from the end of the branch, then interweaving thinner blades of grass into the main frame. Once completed, the nest is subjected to the scrutiny of his chosen partner, who will tear it apart if the result is less than satisfactory!

Many weavers nest in colonies. Their nests, perhaps relying on safety in numbers, are generally a relatively plain oval in design, with an unadorned entrance hole. The nests of more solitary weavers tend to be more elaborate, protecting against egg-eating invaders by attaching a tubular entrance tunnel to the base – in the case of the spectacled weaver (*Ploceus ocularis*), one twice as long as the nest itself. Much scruffier are the nests built by the seven species of sparrow- and buffalo-weaver, relatively drab but highly gregarious dry-country birds that occur throughout drier parts of the region. By contrast, the water-associated thick-billed weaver (*Amblyospiza albifrons*) constructs a distinctive domed nest which is supported by a pair of reeds, and woven as precisely as the finest basketwork.

Waxbills and whydahs

Represented by around 50 species in East Africa, members of Estrildidae are small and easily overlooked but often very colourful seedeaters that generally feed on the ground and have distinctive conical (and in some species rather waxy-looking) bills. Among the species more likely to be seen on safari are the gorgeous purple grenadier (*Granatina ianthinogaster*), the handsome red-cheeked cordon-bleu (*Uraeginthus bengalus*), the psychedelic green-winged pytilia (*Pytilia melba*) and the rather plainer common waxbill (*Estrilda astrild*). The ubiquitous African firefinch (*Lagonosticta rubricata*) is a bright red waxbill that often frequents human habitation, while the three mannikin species (genus *Spermestes*) are brightly marked black-and-white and chestnut finches found in wetlands, forest fringes and suburban

gardens. Arguably the prettiest of all waxbills, however, are the forest-dwelling twinspots and bluebills, none more so that the gorgeous red-headed bluebill (*Spermophaga ruficapilla*), which is often seen in the vicinity of Uganda's Bwindi Impenetrable and Kibale national parks.

Closely related to the waxbills, to which they are brood-parasites in the manner of cuckoos, the whydahs (family Viduidae) are colourful savannah birds most notable for the oddly shaped and very long tails – in some cases up to four times the body length – that males sport in breeding season.

Sparrows

The familiar small seedeaters of the family Passeridae are represented in East Africa by ten indigenous species, most of which are rather nondescript and unobtrusive. The introduced European house sparrow (*Passer domesticus*) has become very common in several urban centres, Nairobi included, since its probably accidental introduction along the coast during the 1960s.

Canaries and buntings

Also known as seedeaters or serins, the canaries of East Africa are mostly quite unobtrusive, with busy but unmemorable songs, and many of the region's species are streaky brown rather than yellow in overall coloration. At least 18 species are present, the most common being the pretty yellow-fronted canary (*Crithagra mozambica*), though this is replaced by the similar white-bellied canary (*C. dorsostriatus*) in drier parts of the interior. Loosely related to canaries, buntings are small but handsome ground-dwelling seedeaters with characteristic black-streaked head patterns. The golden-breasted bunting (*Emberiza flaviventris*) is common in several East African game reserves, while the cinnamon-breasted rock bunting (*E. tahapisi*) is a widespread savannah species that is often but not exclusively associated with rocky slopes and cliffs.

Top Female purple grenadier (AVZ)
Centre Kenya rufous sparrow (*Passer rufocinctus*) (AVZ)
Below The golden-breasted bunting is a small but brightly coloured resident of moist savannah and rocky slopes. (AVZ)

REPTILES AND AMPHIBIANS

Colourful agama lizard, Taita Hills Wildlife Sanctuary (AVZ)

REPTILES

From fire-breathing dragons to the death-eyed basilisks of Greek mythology and apple-pushing serpent of Eden, reptiles have long garnered more than their fair share of bad press. In a few cases this fearsome reputation is not entirely without foundation, particularly with respect to the rapacious Nile crocodile, which kills dozens of African villagers annually. Otherwise, however, the vast majority of East Africa's almost 500 reptile species pose no threat whatsoever to humans – the main exception being a few species of venomous snake, but even these tend to be slow on the offensive. And for those who believe that the only good snake is a dead snake, the ecological value of reptiles cannot be overstated: a healthy snake population helps prevent plague-like outbreaks of rodents; the lizards that skid around hotel walls do much to control mosquitoes; and even crocodiles can be viewed as the aquatic equivalent to vultures, devouring the carrion that might otherwise clog up and pollute East Africa's lakes and rivers.

Unlike mammals or birds, the group of creatures traditionally defined as reptiles does not form a valid clade (a term used to denote the full set of organisms with any given common ancestor) and no single evolutionary adaptation (such as hair or feathers) can be regarded as universally and uniquely reptilian. Indeed, a combination of fossil evidence and comparative DNA testing indicates that crocodiles and turtles are less closely affiliated to lizards or snakes than to birds (which are now known to have evolved from dinosaurs). As a result, the class Reptilia is now split into two subclasses: Lepidosauria, which includes lizards and snakes; and Archelosauria, which incorporates turtles and crocodiles, as well as dinosaurs and birds.

Despite this, most reptiles have horny epidermal scales and paired limbs with five toes, and all breathe through lungs instead of gills and have internal fertilisation. Like amphibians and fish, most reptiles are primarily ectothermic ('cold-blooded'), which means that they regulate their body heat using external rather than internal sources – often by basking in the sun or lying in the shade. For this reason, reptiles are generally most abundant in warm climates and tend to be poorly represented at high altitude. Modern reptiles tend to be relatively small, at least when compared with extinct cousins such as the 150-tonne aquatic *Liopleurodon* (the largest land predators that ever lived) and 70-tonne sauropods (the largest ever terrestrial animals), which thrived during the so-called 'Age of the Dinosaurs' up until 65 million years ago.

CROCODILES

With their armour-like scales, fearsomely toothy mouths and long powerful tails, crocodiles are decidedly sinister and prehistoric in appearance, and it comes as no surprise to learn that, together with birds, they are the last surviving descendents of the gigantic predatory archosaurs that once roamed the earth. Found primarily in tropical freshwater habitats, these large aquatic carnivores are placed in the order Crocodilia, along with the New World alligators and caimans. This order dates back almost 100 million years, and is more closely related to birds and dinosaurs than to other living reptiles – as can be seen from the four-chambered heart, which resembles that of birds and mammals. Indeed, fossilised crocodiles that

A Nile crocodile, *Crocodylus niloticus*, basking on the Rufiji River in Nyerere National Park, displays the fearsome teeth it uses to drag its prey underwater (*above*, AVZ). It also represents one of the most successful carnivorous lines in evolutionary history, with an overall design that is little changed since the age of the dinosaurs (*below*, AVZ).

lived contemporaneously with dinosaurs are sufficiently similar to their modern descendants to be placed in the same family by taxonomists.

Three of the world's 14 crocodile species have been recorded in East Africa, with the Nile crocodile (*Crocodylus niloticus*) being by far the most common. This is also Africa's bulkiest and longest-lived predator, typically reaching a length of 5m, weighing up to 1,000kg, and with a lifespan comparable to humans. The largest individual ever measured is a 7.9m giant shot by hunters in Uganda, but (despite frequent claims to the contrary) the accolade of 'world's largest living reptile' is more properly bestowed on the Australasian estuarine crocodile, which routinely attains a length of 7m – and even this pales in comparison with the extinct 12m-long crocodilians of the genus *Phobosuchus*.

The Nile crocodile is widespread in East Africa, occurring naturally in most large rivers and lakes below an altitude of around 1,600m. Unprotected populations have declined greatly in recent decades, and several million individuals have been killed by professional skin-hunters and vengeful local villagers since the 1960s, but the species remains very common in many national parks and other reserves, and is the least threatened of the world's crocodilians. Lake Turkana, in northern Kenya, harbours the region's densest population, with an estimated 10,000–15,000 individuals regularly breeding on one of its islands, but the Rukwa Basin in southern Tanzania is also known for its prodigious crocs. More accessible to the average tourist are the Nile below Murchison Falls and the Rufiji River in Selous Game Reserve, both of which are known for the gargantuan specimens that bask toothily on the sandbanks – a truly primeval and chilling sight, as they slip silently into the water and vanish at the approach of a boat.

Crocodiles devote more energy to raising their young than do most reptiles. The female lays up to 100 (but more typically 30–50) hard-shelled eggs by night in

Basking crocodiles alongside Tanzania's Rufiji River retreat to the safety of the water when disturbed. (AVZ)

Like other ectothermic creatures, crocodiles derive much of their body heat from basking in the sun. (AVZ)

a small hole that she digs in the sand using her hind legs. She then covers up the hole to hide it from predators. Some females guard the nest vigilantly, but most will abandon it, only returning about three months later to listen for the high-pitched call made by the hatchlings as they emerge from the eggs. The female then carries the hatchlings in her mouth to the water, where she leaves them to fend for themselves. Unlike adults, which have no natural enemies other than man, hatchlings fall prey to a variety of carnivores and their mortality rates can be as high as 98%. The sex of a crocodile depends largely on the incubation temperature: the brood will be predominantly male if this is above 30°C and mostly female if it's lower.

AMBUSH

A Nile crocodile feeding on a blue wildebeest it drowned during a Mara river crossing in the north of Serengeti National Park, Tanzania. (AVZ)

Nile crocodiles feed mainly on fish, occasionally supplementing this staple diet with aquatic birds, terrapins, other small swimming creatures and submerged carrion. Less often and more opportunistically, they will drag a drinking or swimming mammal under water until it drowns, then tear it to pieces, rotating on their axis as they strip away chunks of flesh with their sharp teeth. A large crocodile is capable of killing a lion or wildebeest – or human being, for that matter – and crocodiles have been known to attack animals as large as adult rhino or giraffe. In certain areas, for instance the Mara and Grumeti rivers in the Serengeti-Mara ecosystem, mammals form the main prey of crocodiles – as seen to grisly but spectacular effect when the wildebeest migration undertakes the river crossings.

On a more pragmatic note, river-dwelling (as opposed to lake-dwelling) crocodiles are said to be responsible for the majority of attacks on humans in East Africa, possibly because river water tends to be muddier while lakes provide better visibility for hunting fish under water. Either way, unless you have reliable local information to the contrary, assume that bathing in any lake or river is unsafe, bearing in mind that a croc can submerge without drawing breath for 45–60 minutes and might easily be mistaken for a floating log when it does surface. Crocodiles seldom attack outside their normal aquatic hunting environment, so you are at no appreciable risk if you keep a metre or so back from the water's edge.

Also recorded in the region are two harmless (to humans, anyway) species whose ranges extend into the western Rift Valley. The Central African slender-snouted crocodile (*Mecistops leptorhynchus*), which can grow to be 3m long and has a diagnostic tapering snout, is a shy resident of Lake Tanganyika and possibly a few associated streams, but is unknown elsewhere in East Africa. The world's smallest crocodilian, growing to a maximum of 1.2m, Osborn's dwarf crocodile (*Osteolaemus osborni*) has been recorded in a handful of forested localities in western Uganda, with the most recent substantiated sighting dating to the 1940s. None of the world's eight alligator species occurs in Africa.

TORTOISES AND ALLIES

Tortoises, terrapins and turtles (known collectively as chelonians or shield reptiles) are nature's original backpackers and caravanners, carrying their home – or should that be fortress? – with them wherever they dawdle. That home, of course, consists of a practically unbreakable armoured exoskeleton or shell, which protects them against most predators, so that the small proportion of individuals that survive to adulthood can enjoy a relatively sedate and untroubled life thereafter. Terrestrial chelonians are referred to as tortoises, while their freshwater and marine cousins are called terrapins and turtles respectively. All are placed in the order Testudines, a lineage that first appears on the fossil record more than 200 million years ago. Roughly 350 species have been described worldwide, of which six tortoises, 12 terrapins and five turtles occur in East Africa. Where other reptiles are almost exclusively carnivorous, most chelonians are omnivorous: the terrestrial tortoises tend to have a predominantly vegetarian diet, while terrapins subsist mainly on small aquatic creatures and carrion, but many species will also consume plant matter. Globally, many chelonians are under threat as a direct result of human activity: most species of marine turtle

The leopard tortoise is the largest tortoise in East Africa. (AVZ)

are officially listed as endangered, and many wild tortoise populations are also in decline – particularly in Asia, where they are routinely collected for food, medicine and the pet trade.

Tortoises

The most visible chelonian on safari is probably the leopard tortoise (*Stigmochelys pardalis*), which can be distinguished from other terrestrial species by its large size (adults are 35–70cm in length and can weigh up to 40kg) and the tall domed gold-and-black mottled shell alluded to in its name. Known to live for more than 50 years in captivity, adults have few natural enemies other than humans, but their lack of mobility make them highly susceptible to bush fires. Associated with semi-desert and savannah habitats below 1,500m, this species is often seen motoring in the slow lane of game reserve roads in Kenya and northern Tanzania, but is absent from Uganda and elsewhere west of Lake Victoria.

Also present in East Africa are four hinged tortoises of the genus *Kinixys*, all of which are considerably smaller than the leopard tortoise, and have a pair of fibrous hinges at the back of the shell that allows them to withdraw their hind parts completely while they hide headfirst in a burrow or hole. The peculiar pancake tortoise (*Malacochersus tornieri*), placed in a monotypic genus, is a small (12–18cm) brown East African endemic that has a softer, flatter shell than any other local species. Associated with rocky outcrops in low-lying savannah in Kenya and northern Tanzania, it is known for its ability to climb very steep rocks and other obstructions (including chicken-wire fences), though it spends much of its time tucked safely away in narrow rock fissures or below boulders.

Terrapins

East Africa's dozen or so freshwater terrapins are generally flatter and plainer brown than tortoises, and are usually seen in or close to water, sunning on partially submerged rocks or logs, or peering out from roadside puddles. The Nile soft-shelled

Marsh terrapins are frequently seen basking on rocks alongside the freshwater pools and lakes they inhabit. (AVZ)

terrapin (*Trionyx triunguis*) is the largest species in the region, and is most likely to be seen around Lake Turkana or on the Nile below Murchison Falls, where it can be distinguished by its large size (up to almost 1m) and very wide shell. Nearly as large, the Zambezi flap-shelled terrapin (*Cycloderma frenatum*) is associated with river systems such as the Rufiji and Rovuma in southern Tanzania.

At least nine hinged terrapin species of the genus *Pelusios* occur in the region, but most have a rather localised distribution, the main exceptions being the medium-sized yellow-bellied hinged terrapin (*P. castanoides*) of the coastal belt, and the much larger (up to 55cm) serrated hinged terrapin (*P. sinuatus*), which occurs along most large river systems in eastern Kenya and Tanzania, as well as on lakes Tanganyika and Turkana, and is often seen at the hippo pools in Nairobi National Park.

The smaller marsh terrapin (*Pelomedusa subrufa*) inhabits waterholes, puddles and other stagnant bodies in savannah habitats throughout most non-arid parts of East Africa below 1,500m. It often wanders considerable distances between suitable pools in rainy weather, and will usually aestivate during the dry season, burying itself deep in mud or sand and then re-emerging after the first rains as if from nowhere – giving rise to a local legend that terrapins drop from the sky during storms.

SNAKES

The spotted bush snake, Philothamnus semivariegatus, is an alert non-venomous snake. (AVZ)

Snakes were revered by the ancient Egyptians and their flesh is regarded to have elixir-like properties in certain Asian societies, but in most other cultures they are among the most reviled of all wild creatures and often killed indiscriminately as a result. Snakes are well represented in East Africa, with an estimated 200 species recorded, but they wisely shy away from human contact, and are seldom seen unless actively sought. Most species are legless, although some species still possess tiny vestigial limbs, and a very long individual might possess more than 300 ribs. Snakes smell with their flickering forked tongues and lack external ears, though they are highly sensitive to seismic vibrations, which often warn them of approaching danger. All snakes are carnivorous, feeding on live prey, which they swallow whole. Many play an important role in controlling rodent populations – indeed the persecution of snakes by farmers in parts of Asia has led to several destructive outbreaks of rats since the 1980s. Some also often take eggs and nestlings, for which reason a big commotion among birds in a tree is often a good indication that a snake is around.

Despite their bad reputation, snakes rank far lower on the list of potential African safari hazards than many people suppose. Most are completely harmless.

The African rock python (*above*) and puff adder (*below*) are among the most widespread and commonly observed snakes in East Africa. The former, despite its imposing bulk, is a non-venomous snake that poses no real threat to adults. The latter, however, is responsible for a high proportion of serious snakebites in the region, thanks to its disinclination to beat a retreat when approached by people. (AVZ)

198

Of the 45 venomous species known from East Africa, only 18 are on record as having caused a human fatality, and even these will normally slither away unseen if they hear a human approach. No reliable figures are available for East Africa, but it is probable that the annual tally of snakebite deaths is far lower than the number of people killed daily on the roads, or by malaria. In South Africa, a country with 14 deadly snakes, you're statistically more likely to be killed in a lightning strike.

Pythons

The rock python is Africa's largest snake and one of the more likely species to be seen on safari. It is known to reach lengths of 5–6m (three times as long as a person), through unsubstantiated records of 7.5m and 9.8m individuals have been reported from west Africa. Non-venomous, pythons kill their prey by strangulation, wrapping their muscular bodies around the victim until it can no longer breath, then swallowing it whole and dozing off for a couple of months while the protracted digestive processes kick in. Pythons feed mainly on small antelope, large rodents and similar prey. They are harmless to adult humans, but could conceivably kill a small child. A python might be encountered almost anywhere in the region – typically basking on rocks by day or crossing a road at night – and the Serengeti and Queen Elizabeth national parks are both good spots. Some authorities recognise two species: the southern African rock python (*Python natalensis*), which is mottled gold, brown and grey, and occurs widely throughout Tanzania and south-central Kenya; and the darker, more richly coloured central African rock python (*P. sebae*), which occurs in Uganda, the Lake Victoria basin, the coast north of Tanga and the Kilombero Valley in southern Tanzania.

Vipers

The vipers and adders of the family Viperidae are widespread and often highly venomous snakes characterised by a broad triangular head, thickset build, long hinged fangs and in most cases a nocturnal hunting pattern. Conventionally, the name 'adder' applies strictly to those members of the family that give live birth, while 'viper' refers to egg-layers, but this distinction is rendered meaningless in East Africa by several anomalies – for instance, that all the region's night adder species (genus *Causus*) do lay eggs, while the Gaboon viper doesn't.

The most commonly encountered of the region's 21 species is the puff adder (*Bitis arietans*), a large, thickset resident of savannah and rocky habitats, with the triangular, well-defined head characteristic of all vipers. Although this species feeds mainly on rodents, it will strike when threatened and is considered the most dangerous of African snakes – not because it is especially venomous or aggressive, but because its notoriously sluggish disposition means it is more often disturbed than other snakes. The Gaboon viper (*B. gabonica*) is the largest African viper, attaining lengths of up to 2m. It is very heavily built, and has a beautiful geometric skin pattern in gold, black and brown that blends perfectly into the forest-floor leaf-litter it inhabits. Although highly venomous, this species is both more placid and more elusive than the puff adder.

Cobras

Among the most dangerous snakes are the cobras of the genus *Naja*, some of which can reach a length of 2.5m or even 3m. All species have a trademark hood that they expand as a warning when threatened, raising their head at the same time. The two species of spitting cobra have more diluted venom than other species, which they can spit in self-defence through holes in the front of their fangs, usually aiming at the eyes of their victim. This causes temporary blindness and intense pain, but is unlikely to do long-term damage if you wash the eye out immediately – ideally with clean water, though any liquid will do the trick if need be, including milk or even urine. Many cobras' first line of defence, however, is neither to spit nor to strike but to sham death, which they can do very convincingly – so under no circumstance should you touch or approach what might appear to be a dead snake, but give it as wide a berth as possible.

Green mamba (SSp)

Mambas

Another widespread family is the mambas, of which the black mamba (*Dendroaspis polylepis*) is the largest venomous snake in Africa, measuring up to 3.5m long, and can be distinguished by its narrow head, which is often (and rather aptly) described as coffin-shaped. Legendarily fast-moving, mambas have a reputation for unprovoked aggression, especially breeding males, but there is little evidence to back this up. Like most other snakes, they will attack only when cornered. Mamba bites are rare but should be taken very seriously: the venom, both neurotoxic and cardiotoxic, is very fast-acting, with a real risk of death (often linked to respiratory failure) within 24 hours of an untreated bite. Generally paler and greener in overall coloration, the eastern green mamba (*D. angusticeps*) and Jameson's mamba (*D. jamesoni*) are also highly venomous, but both species are localised in East Africa, and are more arboreal in habits and retiring by nature than the black mamba.

Mozambique spitting cobra (SS)

FACT OR FICTION?

Snakes have been enshrouded in folklore and superstition since biblical times, and even today they inspire a litany of misinformation – from the belief that certain species can hypnotise their prey or inject poison through their flickering forked tongue, to the notion that they habitually swallow their young. But no African species inspires quite such a volume of legend as the black mamba, which is widely characterised as the most intelligent, fearless, swift and vengeful of serpents. One myth has it that the black mamba can roll itself into a hoop and trundle downhill in pursuit of its prey, straightening up like an arrow when it comes within striking range. Other popular accounts have this feared serpent balancing upright on the tip of its tail, moving so quickly that it creates a whirlwind, overtaking people on horseback, milking cattle when it runs out of food, and reconstituting itself after being cut into small pieces. Some even say that the black mamba can ambush the car of an enemy by coiling itself around the wheel of the moving vehicle, and biting the driver when he stops to disembark. All nonsense, of course, for while this large venomous snake will certainly stand its ground when cornered, it is far more likely to flee human contact at the slightest opportunity. Sadly, however, the main victims of widespread mamba-phobia are other more benevolent large brownish species such as the mole snake or brown house snake, both of which are frequently misidentified as mambas and killed.

Vine snakes are extremely hard to detect in the tree canopy. (AVZ)

Though highly venomous, the boomslang is back-fanged and therefore cannot easily bite people. (SSp)

'Typical' snakes

The majority of East Africa's snakes are placed in Colubridae, a family of so-called 'typical snakes', most of which are non-specialised hunters that lack venomous fangs and are harmless to any living creature much bigger than a rat. An exception to this is the boomslang (*Dispholidus typus*), which – as suggested by its Afrikaans name which means, literally, tree-snake – is a largely arboreal snake of moist woodland and savannah. Variable in coloration (most often green but also brown or olive), the boomslang is theoretically the most toxic of Africa's snakes, but because it is back-fanged and very docile, it seldom bites people unless they attempt to handle or catch it. Other back-fanged snakes that pose little threat to humans in the wild are the twig (or vine) snakes (*Thelotornis* spp.), named after their habit of sitting motionless in a tree, head facing outward like a twig, waiting to strike any passing prey. Like the boomslang, their fangs are set too far back to represent a serious threat to a casual passer-by: one would practically have to chew on you in order to inject its slow-acting venom!

Probably the most common snake in East Africa, the brown house snake (*Boaedon fuliginosus*) is highly variable in colour, ranging from olive-green or brown to orange and sometimes even black, and tolerates a variety of habitats. It often occurs in suburban gardens and in other low-rise urban environments, where it plays a valuable role in controlling rats and other pests. The mole snake (*Pseudaspis cana*) is a common grey-brown savannah resident that grows up to 2m long, and feeds on rodents in their underground burrows, making it a valuable ally to farmers.

The brown house snake is essentially harmless to people and preys mainly on rats, mice and other potential household pests. (AVZ)

Other regularly observed colubrids include the dozen or so 'green snake' species of the genus *Philothamnus*, which are generally bright green with large dark eyes, and are often seen near water, where they feed on amphibians. Most common and widespread East Africa is the fast-moving and predominantly arboreal spotted green (or bush) snake (*P. semivariegatus*), which typically has green upper scaling marked by bold black bands, grows to be more than a metre in length, and whose eyes are shielded by a distinctive brow ridge. Other commonly seen species include Battersby's green snake (*P. battersabyi*) of the Tanzania/Kenya border area, and the speckled green snake (*P. punctatus*) of southern Tanzania.

A widespread and quite remarkable colubrid, the rhombic egg-eater (*Dasypeltis scabra*) lives exclusively on bird's eggs, dislocating its jaws to swallow the egg whole, then eventually regurgitating the crushed shell in a neat little package. Its complex markings are very similar to those of the night adders, and it is thought that this is an example of Batesian mimicry, whereby a harmless animal has evolved to resemble its venomous counterpart in order to gain a measure of protection by deceiving would-be predators.

The speckled green snake is a pretty coastal species with distinctive large eyes. (AVZ)

LIZARDS

The most numerous and conspicuous of East Africa's reptiles, lizards are classified alongside snakes in the order Squamata, though they inspire far less fear than their legless serpentine cousins. Among more than 200 species recorded in the region, none is venomous, nor can any be regarded as a threat to humans – though the hefty monitor lizards might, in theory, inflict a nasty bite if cornered. Visitors to East Africa are almost certain to see lizards during the course of a safari, and should not be concerned at the presence of a few individuals in their hotel room. On the contrary, these are wholly innocuous and might even be considered desirable roommates, since they tend to snaffle up mosquitoes and other insects.

Monitors

One lizard you probably wouldn't want to find in your room is a monitor of the genus *Varanus*, which is represented by four species in East Africa – though two are very rare and localised within the region. Most likely to be seen is the Nile monitor (*V. niloticus*), which is Africa's largest lizard, regularly reaching a length of 2.5m and sometimes more. A meat- and carrion-eater, it is dull green-brown in colour – usually with some yellow banding on the torso, which fades with age – and is common in the vicinity of rivers, lakes and most other aquatic habitats below altitudes of around 1,600m, where it is sometimes mistaken for a young crocodile. The sandy or grey-brown rock (or white-throated) monitor (*V. albigularis*) has a blunter nose and a shorter tail, and is generally seen away from water – often in the vicinity of termite mounds in drier savannah habitats in Kenya and Tanzania.

The Nile monitor is usually seen close to water, either basking on the rocks or crashing through the undergrowth. (AVZ)

Top Red-headed rock agama (AVZ)

Above Tropical house gecko (SSp)

Below The speckle-lipped skink (*Mabuya maculilabris*) is a localised lizard named for the black and white speckling around its mouth. (AVZ)

Agamas

Also very common and conspicuous, the eight species of the genus *Agama* are distinguished from other common lizards by their relatively large size of around 20–30cm, their big head and their almost plastic-looking scaling – depending on the species, a garish combination of blue, purple, orange or red, with the head generally a different colour from the torso. Particularly spectacular are the males of the red-headed rock agama (*A. agama*) and Mwanza flat-headed agama (*A. mwanzae*), with the latter commonly observed basking on rocks in the Serengeti-Mara ecosystem, and often very confiding and easily photographed in rocky lodge grounds.

Geckos

The most diverse lizard family in East Africa is the geckos, represented by at least 55 species, most of which are rather small, plump and pale, with large lidless non-closable eyes, and unique adhesive toes that enable some species to run upside down on smooth surfaces. Safari-goers will soon become familiar with the tropical house gecko (*Hemidactylus mabouia*), an endearing bug-eyed, translucent white lizard, which as its name suggests reliably inhabits most houses as well as lodge rooms throughout the region, scampering up walls and upside down on the ceiling in pursuit of pesky insects attracted to the lights. Several species of gecko are endemic to the Eastern Arc Mountains of Tanzania, the most notable perhaps being the spectacularly coloured turquoise dwarf gecko (*Lygodactylus williamsi*), which has only ever been observed in the Mimboza Forest near Morogoro.

Skinks and others

Another common family is the skinks: a group of small, sleek, long-tailed lizards represented in East Africa by 45 species, of which the most widespread and visible are the variable skink (*Trachylepis varia*), striped skink (*T. striata*), five-lined skink (*T. quinquetaeniata*) and rainbow skink (*T. margaritifera*), all of which are quite dark and slender, and have a few thin stripes running from head to tail. East Africa is also home to 19 members of Lacertidae – a variable Old World family sometimes referred to as the true lizards. Perhaps the most remarkable East African member of this family is the blue-tailed gliding lizard (*Holaspis guentheri*), a striped blue coastal forest species whose unusually flattened body and tail allow it to glide between trees for distances of up to 10m.

Chameleons

Arguably the most charismatic and intriguing of African reptiles, true chameleons of the family Chamaeleondae are confined to the Old World, with the most important centre of speciation being Madagascar, to which about half the world's 130 species are endemic. Another two species of chameleon occur in each of Asia and Europe, while the remainder are distributed across mainland Africa. Madagascar aside, East Africa supports the world's greatest chameleon diversity, with 11 Tanzanian endemics, four Kenyan endemics, five cross-border East African endemics and three Albertine Rift endemics listed among the region's 40 described species. The

Flap-necked chameleon (AVZ)

name chameleon derives from the ancient Greek *chamae leo*, meaning 'dwarf lion', a reference to this lizard's intimidating defensive behaviour, which involves inflating its body, opening its mouth wide, and hissing as it lunges forward.

Chameleons are known for their abrupt colour changes, a trait that's been exaggerated in popular literature, and is influenced more by mood than background colour. Most African chameleons are green, taking on a browner hue when they descend to more exposed terrain, but undergoing dramatic colour and pattern changes only when threatened. Different chameleons vary greatly in size. Quite common in coastal forest and low-lying woodland in eastern Tanzania, Meller's (or giant one-horned) chameleon (*Trioceros melleri*) is the largest African species, but its maximum length of 55cm is outdone by the 70cm-long Oustalet's chameleon (*Furcifer oustaleti*) of Madagascar.

A remarkable physiological feature common to all chameleons is their protuberant round eyes, which offer 180° vision on both sides and can swivel independently of one another. Only when one eye isolates a suitably juicy-looking insect will both focus in the same direction, as the chameleon stalks slowly forward until it is close enough to use the other unique weapon in its armoury – its huge sticky-tipped tongue. Typically about the same length as its body, the tongue remains coiled within the mouth most of the time, to be unleashed in a sudden, blink-and-you'll-miss-it lunge at a selected item of prey. In addition to their unique eyes and tongues, many chameleons are adorned with facial casques, flaps, horns and crests, which enhance their already somewhat fearsome prehistoric appearance.

You'll most likely come across a chameleon when it is crossing the road, in which case it should be easy to take a close look, since it moves painfully slowly. The flap-necked chameleon (*Chamaeleo dilepis*) is the most regularly observed savannah

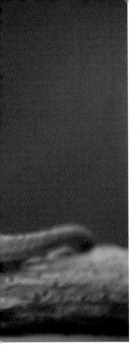

and woodland species in Tanzania and southern Kenya, though it is replaced by the graceful chameleon (*C. gracilis*) in Uganda and parts of Kenya. Both species are typically 15–20cm long, and bright green in colour with few distinctive markings. Chameleons are often seen on night game drives, when their white nocturnal colouring makes them easy to pick up by spotlight. This spectral appearance is just one reason why chameleons are feared by most rural Africans, who will generally refuse to touch one. It is widely held that the chameleon is a harbinger of a death in the family, and several East African cultures blame its ponderous locomotive style for the fact of death itself – the story being that a chameleon was sent by god to tell men that they would no longer die, but he was beaten by a bird, who delivered the opposite message.

The ancient forests of Tanzania's Eastern Arc Mountains and southern highlands harbour a wealth of curiosities, notably the tubercle-nosed chameleon (*T. tempeli*), the spiny-flanked chameleon (*T. laterispinis*), the Pinocchio-like sharp-nosed chameleon (*Kinyongia oxyrhina*) and the tiny (5cm) Uluguru pygmy chameleon (*Rhampholeon uluguruensis*). Several of these Tanzanian endemics are very recently described, and it seems certain that further species await discovery in this biologically exciting region.

The West Usambara two-horned chameleon (*Kinyongia multituberculata*) is endemic to the Usambara mountains of northern Tanzania. (AVZ)

The beautifully marked banded rubber frog (*Phrynomantis bifasciatus*) is a widespread species that might be seen in any freshwater habitat. (AVZ)

AMPHIBIANS

Put simplistically, amphibians form the 'missing link' between fish and amniotic vertebrates, having evolved almost 400 million years ago, thus pre-dating the arrival of reptiles on the fossil record by around 100 million years. Most of the world's 8,000 species are partially terrestrial but depend on fresh or brackish water in which to breed, starting life with exterior fish-like gills but later metamorphosing into a distinct adult form that can breathe through its skin and has rudimentary lungs. The most diverse of the world's three extant amphibian orders is Anura (frogs), with at least 7,000 species described to date. More than 550 salamander species of the order Caudata are recognised, but none occurs south of the Sahara, while East Africa is home to nine of the world's 170 species of caecilian, a group of small earthworm-like creatures placed in the class Gymnophiona.

FROGS AND TOADS

East Africa is known to harbour at least 200 frog species but probably has many more, with new ones being discovered and described on a regular basis, particularly in isolated forest regions such as the Eastern Arc and Albertine Rift. Most are more or less nocturnal, spending the daylight hours under water or in cool, moist places (such as burrows, hollows or leaf-litter) to prevent their permeable skin from drying out. As is the case with many birds, male frogs attract potential partners with mating calls. These are unique to each species and often form the best way of reliably telling similar species apart. Calling is generally most vigorous after rain, when males of many species call together to create loud, chaotic choruses, a ploy that gives females coming from afar a more persistent aural target to track down. Because of this, frogs are more often heard than seen, and treat safari-goers to an unforgettable evening medley of guttural croaks and ethereal whistles during the rainy season.

East Africa's frogs are divided into eight families, of which the best known is probably Bufonidae, more commonly known as toads. About 40 toad species are present in the region, most of which are medium-to-large anurids similar in appearance to the familiar European toad. Most toads are highly vociferous in season, with pairs of males often calling back and forth alternately, a phenomenon known as antiphony, and many species are very difficult to tell apart by sight or by call. Probably the most widespread in East Africa is the guttural toad (*Sclerophrys gutturalis*), which is named for its rough croaking call, often performed by dozens of males at the same location. It can grow to be 12cm long, making it the largest toad in the region. Also placed in Bufonidae but at the opposite end of the size spectrum are the dozen 'forest toad' species of the genus *Nectophrynoides*, which is endemic to the mountains of eastern Tanzania.

The largest frog in the region is the African bullfrog (*Pyxicephalus adspersus*), which can reach 24cm long and weigh up to 1kg. This impressively aggressive species aestivates for much of the year, but during the rainy season emerges to breed in seasonal pools in dry savannah. Although people tend to associate frogs with standing water, East Africa's most diverse family is Hyperoliidae, which comprises 60-plus species of smallish and generally distinctively patterned 'tree frogs', most of which are skilled climbers with long broad-tipped toes, and inhabit forest, woodland

African bullfrog (AVZ)

Common reed frog (*Hyperolius viridiflavis*) (SSp)

and reedy habitats. One of the region's most evocative sounds is the ethereal popping chorus of the bubbling kassina (*Kassina senegalensis*), a 'tree frog' of marshes and moist grassland that often collects in vociferous congregations during the late afternoon and early evening, being especially vocal during the breeding season of April to May.

AMPHIBIANS UNDER THREAT

Habitat loss, wetland degradation and industrial pollution have resulted in a sharp global decline in amphibian populations in recent decades. Worldwide, more than 200 species have become extinct since the 1980s, mostly in the Americas and Australasia. East Africa has suffered less than more industrialised and densely populated regions; nevertheless, a quarter of its frog species are on the IUCN Red Data List, with 29 being ranked as Vulnerable, 17 as Endangered and six as Critically Endangered. The region's most endangered species is almost certainly the Kihansi spray toad (*Nectophrynoides asperginis*), which is so tiny that it fits on a human fingernail. It is unusual among amphibians in that it gives birth to a brood of live miniatures. Its range is confined to 1ha of ferny undergrowth sustained by the spray from a single waterfall on the Kihansi River at the base of the Udzungwa Mountains. In 1996, when the species was discovered, the population stood at around 10,000, but the diversion of the river to a hydro-electric dam destroyed 95% of its habitat. It is now extinct in the wild, though captive breeding populations still exist.

INVERTEBRATES

The citrus swallowtail (*Papilio demodocus*) is a large butterfly whose caterpillars feed exclusively on the leaves of citrus trees. (AVZ)

Large mammals and colourful birds tend to hog the safari limelight, but East Africa's wildlife is immeasurably more diverse and prolific at the lower end of the size scale. Here, munching through the vegetation, tunnelling into the ground and swarming through the skies, the engine room of the great food pyramid is powered by the bewildering variety of creepy-crawlies known collectively as invertebrates. In the late 18th century, the term invertebrate (Latin for 'without a backbone') gained taxonomic currency as the name of a phylum that lumped together all animals placed outside the five vertebrate orders (ie: fish, amphibians, reptiles, birds and mammals). It is still in wide use today, embracing some 97% of known animal species, but is no longer regarded to form a true evolutionary clade: a full 30 different invertebrate phyla are now recognised by taxonomists.

The 'lower' invertebrates include soft-bodied annelids such as the burrowing earthworms and blood-sucking leeches, as well as stalk-eyed molluscs, which use a single muscular foot for locomotion. Notable among the latter is the African land snail (*Lissachatina fulica*), whose 15cm shell makes it the world's largest land gastropod. The more advanced arthropods, which have many limbs and segmented bodies, range from simple crustaceans such as woodlice (Isopoda) and freshwater crabs (Decapoda), to the anatomically complex insects and arachnids. Invertebrates being so numerous and varied, coverage here is restricted to a cursory overview of the better- known and more visible groups.

Millipede (RH/AI)

MILLIPEDES AND CENTIPEDES

The different arthropod orders are distinguished from one another by their number of legs, and none has more than the millipedes (Chilognatha), whose body comprises up to 60 segments, with two pairs of legs on each. Millipedes are among the most familiar of East African invertebrates, often seen trundling across roads on a twinkling forest of tiny legs and rolling into a tight ball when disturbed – their popular southern African name *tshongololo* means steam train and is a perfect description of their motion. Often regarded with suspicion, millipedes are harmless vegetarians, but can secrete a noxious fluid in defence and are unpalatable to most creatures – civets and assassin bugs being notable exceptions. Centipedes, by contrast, are carnivores, with only one pair of legs on each of their 25 body segments. The large centipedes (Scolopendromorpha) are the best known of several different groups. These aggressive, fast-moving predators have a painful (though not dangerous) bite, and big specimens – up to 25cm long – will even prey on frogs and geckos. Largely nocturnal, they are also active on overcast days, when their bright orange colouring should be warning enough not to touch. By curling around their young to protect them, female centipedes show more parental care than many vertebrates.

ARACHNIDS

The predatory spiders and scorpions, along with the parasitic ticks and mites, are classified as arachnids, an arthropod class that differs from the superficially similar insects in having a body comprising two (not three) principal sections, eight (not six) legs, and no antennae.

SPIDERS

Spiders can do amazing things with silk, a remarkable substance used to weave webs, make egg sacs, line burrows, lure mates and hitch rides on the breeze. Golden orb spiders (*Nephila* spp.) are the original web masters: the big black-and-yellow females string a concentric cat's cradle of golden rings on a hammock of radiating support

Banded-legged golden orb spider (*Nephila senegalensis*) (AVZ)

wires, then wait at the centre for a catch. Nearby lurks the much smaller male, and often a number of tiny, dew-drop spiders (*Argyrodes* spp.) which scavenge any leftovers. These webs are so strong that they have been known to ensnare swallow-sized birds, and their tensile properties have reputedly been studied by NASA scientists. Kite spiders (*Gasteracantha* spp.), whose spiny, triangular abdomens look like something out of *Star Wars*, also build an orb web – as do bark spiders (*Caerostris* spp.). The latter dismantle theirs by day and retire to a camouflaged position on a nearby branch. Community nest spiders (*Stegodyphus* spp.) nest socially in a large, spherical web, which becomes strewn with debris, while daddy-long-legs spiders (*Smeringopus* spp.) hide from predators by hanging upside down beneath their web and shaking it when disturbed. Other web designs include tunnels built into crevices, sheets spread over the ground and the complex 'scaffold' webs of button spiders (*Latrodectus* spp.), with thread snares that lead to their refuge. Some spiders dispense with a web and actively wield their silk as a weapon: net-casting spiders (*Deinopis* spp.) dangle from low vegetation to fling a silken net over prey passing below, while sand divers (Ammoxenidae) pounce on termites and bind them tightly in silk.

Spiders have perfected many hunting techniques beside the silken snare. Baboon spiders (*Harpactira* spp.) lurk down neatly drilled, silk-lined tunnels, about 3cm across, from which they ambush prey when alerted by its footfalls. Robust, hairy and the size of a small rodent, these impressive tarantulas are the stuff of an arachnophobe's nightmare. Other ambush hunters include trapdoor spiders (Ctenizidae), which construct a cork-like lid to their tunnel. Rain spiders (*Palystes* spp.) are not tied to a web or a tunnel, but hunt their prey by speed – often around houses at night. These large spiders are perfectly harmless, despite the alarm they sometimes cause. Jumping spiders (Salticidae) leap onto their victims, first securing themselves to the ground with a silken line. Fishing spiders (*Nilus* spp.) pursue insects, tadpoles and small fish across the water's surface, using their legs as a lure.

Spiders use fangs to subdue prey with venom. Very few can penetrate human skin, and most of these are not dangerous – although the 'black widow' button spiders (*Latrodectus* spp.) are among a handful whose bite demands urgent medical attention. Ironically the 'spider' with the most impressive fangs is neither a spider nor venomous. Sun spiders, better known as solifuges, belong to the separate order Solifugae. These large, voracious hunters (up to 8cm long) are identified by their red or yellowish colouring, soft hairy body, massive fangs and their habit of dashing around after prey with their front legs held off the ground. Victims, including large beetles and even small frogs, are killed without venom and crunched up audibly. Solifuges are common in hot, dry areas and frequently seen scuttling around by the light of a camp fire.

SCORPIONS

A scorpion keeps its venom in the sting on the end of its tail, which it arches forward over its head to strike prey and intimidate enemies. The fearsome-looking pincers, used to grip and grab, are non-venomous and hold no threat to people. All scorpions can deliver a painful sting, but only those with thin pincers and a thick tail are dangerous, and their potent venom requires urgent medical attention. Some thick-tailed scorpions will also spray their venom in self-defence, and can even produce a warning sound by scraping their sting across their tail. Scorpions with thick pincers and a narrow tail are less venomous. The greatest variety of scorpions occurs in arid areas, where these hardy creatures can survive for a long time without food or water. Being nocturnal, scorpions are seldom seen by day – unless, like a hungry baboon, you are prepared to grub about under rocks for them. At night they venture from their hiding places in search of prey, which they detect on air currents using tiny sensory body hairs. Scorpions are remarkable among invertebrates for their breeding behaviour. Mating partners lock pincers in a grappling courtship pirouette, clearing the ground for the male to deposit his sperm sac, which the female picks up with her abdomen. Newborn young are carried on their mother's back for protection.

The fat-tailed scorpion (*Parabuthus liosoma*) has an extremely potent sting. (SSp)

TICKS

Ticks (Acari) are the strongest disincentive to wearing shorts in the bush, especially during the rainy season in long grass. These tiny parasitic arachnids wait on grass tips and clamber aboard animals that brush past. Once on, they pierce the skin with their sharp mouthparts, and gorge themselves on the host's blood, swelling up like peas. Different species choose different hosts. Cattle and dog ticks (Ixodidae) infest domestic livestock, and the miniscule, red 'pepper ticks' that plague hikers are actually their larvae. Some species of tick can transmit diseases. The common form of 'tick-bite fever' (ricketsia) is not dangerous, but very unpleasant, causing intense headaches and some malaria-like symptoms. Antibiotics will deal with it quickly, but letting the disease run its course – if you can bear it – will produce a level of resistance that usually prevents recurrence. Ticks can be discouraged with insecticide spray, and removed by carefully pinching them out – taking care not to leave the mouthparts embedded. Oxpeckers have honed this skill to a fine art, and with up to 20,000 ticks being recorded on a single giraffe, they're seldom short of a meal. Mites are parasites on larger invertebrates, including spiders, scorpions and millipedes. Most are microscopic, but the larger red velvet mites (*Dinothrombium* spp.) look like miniscule (5mm) scarlet cushions, and can be seen in open, sandy areas in the early morning after rains.

INSECTS

Insects are distinguished from other arthropods by having a three-segmented body (consisting of head, thorax and abdomen), three pairs of jointed legs and – in most cases – two pairs of jointed wings. They are the most diverse higher life form on Earth, with millions of species worldwide, of which tens if not hundreds of thousands are present in East Africa. To many people, insects are simply pests, a description that might fairly (albeit parochially) be applied to the likes of cockroaches, houseflies, head lice, clothes moths, mosquitoes and timber beetles. The bigger picture, however, is that insects play an integral role in the natural environment, whether as pollinators (bees, flies, butterflies), waste disposal machines (dung beetles, termites, blowfly larvae) or providers of such precious natural resources as silk and honey. In East Africa many also form a valuable food source, with locusts, termites and caterpillars all on the menu in various parts of the region.

Insects fall into 30 different orders, with a bewildering variety of families and lifestyles. Some, such as dragonflies and damselflies (Odonata), are predators. These dashing hunters patrol their wetland territories and capture other insects in flight,

Mantids were held sacred by the hunter-gatherer cultures whose prehistoric art adorns rock shelters throughout East Africa. (AVZ)

Crickets are nocturnal grasshopper-like insects whose white noise of chirruping calls is characteristic of East Africa's forests. (AVZ)

using amazing aerial agility and the keen vision of their large compound eyes. The larvae of antlions (Myrmeleontidae) are ambush specialists: they dig a conical pit in loose sand and wait for their ant prey to stumble in, seizing it in powerful jaws and sucking it dry. Praying mantids (Mantodea) hunt using stealth and disguise: some are leaf green, and hide among foliage; others use elaborate camouflage to resemble bark or petals. All have huge eyes, mounted on a mobile, triangular head, and lethal forelegs, held clasped together as if in prayer, with which they snatch victims. In some species, the female chews off her mate's head during mating – though he usually manages to finish the job, since his copulatory movements are regulated by the last ganglion of the abdomen.

Crickets and grasshoppers (Orthoptera) are vegetarians, with powerful hind legs for leaping. They proclaim their territory with a symphony of bleats and chirrups, produced by rubbing various body parts together. Long-horned, nocturnal species are generally known as crickets, while short-horned, diurnal species are generally known as grasshoppers or locusts. Stick insects (Phasmatodea) are also vegetarians, and have a body that exactly mimics a twig or grass stem. Like some mantids and grasshoppers, many flash bold markings on their hind wings to confuse a predator. The largest stick insect species can reach a length of 25cm.

Bugs (Hemiptera) use piercing and sucking mouthparts for extracting juices. There are thousands of species, ranging from predatory assassin bugs (Reduviidae), which can deliver a painful bite from their strong, curved proboscis, to herbivorous stink bugs (Pentatomidae) which suck sap from plants, and cicadas (Cicadidae), whose nymphs feed on root sap below ground, some species taking over 17 years to develop. Adult cicadas live for only a fortnight or so, during which males attract mates with an ear-splitting zinging call created by the rapid vibration of a membrane in their hollow resonating abdomen – one of the most characteristic sounds of the African bush, by day or night. Superficially similar to bugs, cockroaches (Blattodea) are identified by their flattened shape and long, fine antennae. Some exotic species, such as the American cockroach, are much reviled for infesting homes and spreading disease. However, many harmless indigenous varieties can be found living beneath bark or under stones. All cockroaches are omnivorous and most can fly.

Blister beetles (SSp)

BEETLES

Beetles (Coleoptera) constitute by far the largest single order in the animal kingdom, with more than 20,000 species identified in East Africa alone. The sheer variety among beetles is staggering, but all have biting mouthparts and forewings modified into hardened cases (elytra). Some, such as ground beetles (Carabidae) and tiger beetles (Cicindellidae), are fast-moving, wingless predators that hunt down their prey by sight. Others, such as fruit chafers (Cetoniinae), are day-flying vegetarians that feed on flowers, fruit and fermenting sap. Many, including the ominously named blister beetles (Meloidae), pack a noxious defensive spray and sport bright warning colours to advertise the fact. Among the most extraordinary beetles are those of the cantheroid group, better known as glow-worms and fireflies (Lampyridae), which illuminate their abdomen to attract a mate. Most often seen on forest fringes, female glow-worms remain immobile and emit a steady glow, while males emit regular pulses in flight. This is the most efficient form of light known to science, being entirely chemical and generating no heat whatsoever.

Dung beetles (Scarabaeinae) are robust black or brown beetles that can be seen piling into fresh animal droppings and rolling away balls of dung backwards, often crashing into obstacles and battling violently with each other in their haste to secure the prize. It is inside these so-called brood balls that a female dung beetle lays her eggs. One brood ball can be 40 times the weight of its roller, which gives some idea of the beetle's strength.

Dung beetles rolling a ball of dung (AVZ)

Some species roll their brood balls away, others lay their eggs right underneath the mound. Either way, dung beetles play a vital role in soil fertilisation. Their efficiency at waste disposal has even led to them being exported to Australia, which lacks indigenous invertebrates to clear up the countless tonnes of alien cattle dung smothering the outback. Rhinoceros beetles are closely related to dung beetles, and are easily identified by the impressive nose horn – used in territorial combat – from which they get their name.

FLIES AND MOSQUITOES

It is hardly surprising that flies and mosquitoes (Diptera) are among the most unpopular of insects, given their general association with putrefaction and disease. However, like all other 'pests', each species has evolved to suit its environment, and many perform vital ecological services such as waste disposal and pollination. All flies have mouthparts adapted to feeding on liquid matter, which they digest outside the body. There are close to 10,000 species in East Africa, including predators such as robber flies (Asilidae), which snatch other insects in flight, and nectar-eaters such as hoverflies (Syrphidae), which dip into flowers with their long proboscis. Blowflies (Calliphoridae) are carrion-eaters, and lay their eggs in carcasses, from

which their larvae – or maggots – hatch in huge numbers to feast on the rotting flesh. While some flies lap up nectar and juices, others have more carnal tastes. One such bloodsucker, the tsetse fly (family Glossinidae), has proved a historical ally to wildlife by spreading a devastating livestock disease that has impeded the advance of cattle farming. Although its painful bite can transmit sleeping sickness (Trypanosamiasis) and it remains common on many game reserves, the tsetse is not generally considered a serious threat to people.

Unfortunately, the same cannot be said of mosquitoes. These innocuous-looking little flies are primarily nectar-feeders, but females also suck blood

Hoverflies are harmless flies that visit flowers for nectar. (AW/L/AI)

from vertebrate hosts and it is females of the genus *Anopheles* – identified by a body angled downward at rest – that are responsible for transmitting malaria, estimated to kill more than 2 million people in Africa each year. Malaria is caused by a microscopic parasite carried in the mosquito's saliva, and the anaesthetic properties of this saliva leave the victim unaware of being bitten. Since mosquitoes lay their eggs in water, and their larvae are aquatic, they are most common in low-lying, well-watered areas, and that irritating whine in the ear should be taken seriously by any visitor (see *Health and safety*, page 247).

The larger striped swordtail (*Graphium antheus*) is one of more than 1,500 butterfly species in Kenya. All butterflies and moths start life as caterpillars. (AVZ)

BUTTERFLIES AND MOTHS

Butterflies, part of the order Lepidoptera, are the flower children of the invertebrate world, and enjoy a better press than most other insects, including the closely related moths. Actually there is little distinction between the two. Both have tiny, ridged wing scales that create their characteristic bright colour. Broadly speaking, butterflies are colourful, day-flying insects with thin, clubbed antennae, that hold their wings vertically at rest, whereas moths are drab (with some garish exceptions), night-flying insects with un-clubbed antennae, that fold their wings roof-wise. Adults of both groups uncoil a long proboscis to sup on nectar, sap and fruit juices. Their eggs, laid on specific food plants, hatch into voracious caterpillars that develop into pupae from which new adults emerge. Butterfly pupae are naked, whereas moth pupae often wear a silk cocoon.

Anyone brought up on a more limited selection of butterflies – for instance, the UK's 60-odd species, each with a handy common name – is easily dazzled by East Africa's pageant, with 1,200 species having been recorded in Uganda alone, and certain individual forests such as Bwindi or Kibale supporting a checklist of 400-plus. Rotting fruit, fresh dung and evaporating puddles often draw many different species in search of food and mineral salts, while some migrate in countless millions. The most impressive-looking of all butterflies are the 100 species of swallowtail (Papilionidae), which are named for their long tail streamers and are mostly large and colourful – indeed, the African giant swallowtail (*Papilio antimachus*), an endangered west African species whose range extends into Uganda, is often regarded as the world's largest of this family, with a wingspan known to exceed 20cm. Pieridae is a family of medium-sized butterflies, most of which are predominantly white in colour, but with some yellow, orange, black or even red and blue markings on the wings. The diverse family Lycaenidae accounts for a third of Africa's butterfly species, and includes the pretty *Hypolycaena hatita*, a small bluish butterfly with long tail streamers that is often seen on forest paths. The African blue tiger (*Tirumala petiverana*), a large black butterfly with about two dozen blue-white wing spots, is often observed in forest paths near puddles or animal droppings, while the slow-flying African queen (*Danaus chrysippus*) has orange or brown wings and is as common in forest-edge habitats as it is in cultivated fields or suburbia.

If you find butterflies hard to identify, then moths – of which there are ten times as many species – may drive you to despair. A few larger ones are distinctive, such as the spectacular emperor moths (Saturnidae), with their broad wings (up to 15cm across), bold eye markings and – in some species – long tails, and the dashing hawk moths (Sphingidae) with their thickset body and delta-shaped wings. However, many moths are better known by their larvae – or caterpillars. Slug moth caterpillars (Limacodidae) are fat, green and armed with painful, poisonous spines. Tiger moth caterpillars (Arctiidae) are also best left alone, since their coat of woolly hairs contains a nasty toxic irritant. Looper moth caterpillars (Geometridae) take a more discreet approach to self-preservation by disguising themselves as twigs – invisible until they start 'looping' along a branch. Some caterpillars are serious pests: the notorious crop-devouring army worms and stalk borers are the larvae of owl moth species (Noctuidae).

Butterfly pupae at the Kipepeo Project butterfly farm in the Arabuko-Sokoke Forest, Kenya, where they are bred for export to Europe. (AVZ)

Termite mounds may reach an immense height (*main image*, AVZ). After the rains, the winged 'reproductives' (*inset*, SS) disperse to found new colonies.

TERMITES

Termites (Isoptera) are perhaps the most amazing, though in evolutionary terms the most primitive, of the social insects. Popularly known as 'white ants', they are actually unrelated to ants, but have independently evolved many similarities of lifestyle. So successful are termites, that they probably constitute the largest biomass of herbivores in African savannahs (in other words their total weight exceeds that of antelope). All termites share similar social structures. Each colony has soldiers that protect the nest, workers that care for young and forage for food, and an egg-laying queen that is fertilised by a resident king. All termites are vegetarian: some carry micro-organisms in their gut that enable them to digest plant cellulose; others cultivate a fungus (*Termitomyces*) to do this job for them. Although termites can cause serious damage to timber and crops, this is outweighed by the essential work they do in draining, enriching and aerating the soil.

Members of the family Macrotermitidae are responsible for one of the true wonders of nature. Communicating entirely through pheromones, millions of blind worker termites can raise several tonnes of soil, particle by particle, into an enormous structure – over 3m high in some species – in which separate chambers house brood galleries, food stores, fungus combs and the queen's royal cell. The long-lived queen produces up to 30,000 eggs a day, which means that the millions of inhabitants of the colony are all brothers and sisters. The whole structure is prevented from overheating by a miraculous air-conditioning system. Warm air rises from the nest chambers, up a central chimney, into thin-walled ventilation flues near the surface (you can feel the warmth by placing your hand inside one of the upper vents). Here it is cooled and replenished with oxygen, before circulating back down through separate passages and cavities into the nest chambers, passing en route through specially constructed cooling vanes, kept damp by the termites. In this way, termites maintain the 100% humidity and constant temperature of 29–31°C required for successful production of eggs and young. After the rains, when conditions are right, the queen produces a reproductive caste of winged males and females – known as imagoes – that leaves the colonies in huge 'emergences' to mate, disperse and establish new nests, providing a seasonal feast for many other creatures.

ANTS, WASPS AND BEES

Most ants, bees and wasps are social insects, though they are placed in a more advanced order (Hymenoptera) than termites, with all species passing through a larval and pupal stage. Ants (Formicidae) are the most advanced of all. Some species live in underground colonies of millions. Others live in smaller colonies inside hollow acacia thorns or use chewed plant matter to build their nests in trees. All colonies contain one queen and many wingless workers, some of which are soldiers responsible for defence. When conditions are right, the queen produces winged males and females that leave the nest in mass nuptial flights. Males die immediately after mating, but females disperse in search of new nest sites. Ants feed in different ways. Some, such as driver ants (Dorylinae), swarm over large areas in a relentless quest for food, devouring any prey unable to escape (and ruining many a promising picnic). Others, such as myrmicine ants, 'farm' aphids inside their nests, stimulating them to produce 'honey dew' (excreted plant sap) for their larvae. Communication within the colony involves a combination of scent trails, pheromones and, in some species, chirping calls. Threats are repelled with bites or stings, and formicine ants (Formicinae) can also spray formic acid at an attacker, leaving a distinct vinegary smell.

Closely related to ants, wasps are day-flying insects, with two pairs of wings and usually a thin waist between the thorax and abdomen. The females of many species use their modified ovipositor to inflict a powerful sting in self-defence. This

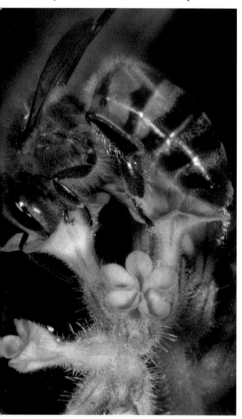

lethal weapon is also used by parasitic species, such as spider-hunting wasps (Pompilidae), to paralyse prey, which they feed to their carnivorous young. Some parasitic wasps cache their victim in a burrow and lay their eggs on it, so that the newly hatched larvae have a ready, live food source. Ichneumon wasps (Ichneumonidae) use their long ovipositors to burrow into plant stems and lay their eggs directly into the larvae of wood-boring moths and beetles. Other wasps construct nests for their young: potter wasps (Eumenidae) create a variety of ingenious chimneys and pots out of mud; paper wasps (Vespidae) work with wood pulp to build delicate multi-celled apartment structures, which they defend against all comers.

Unlike most wasps, bees (Apoidea) feed their young on nectar and pollen, helping to pollinate plants in the process. Some varieties are solitary, such as the hole-drilling carpenter bees (Xylocopinae), which resemble European bumblebees. Others, such as the honeybee (*Apis mellifera*), are highly colonial.

Honeybee (RH/AI)

A honeybee colony, often located in a tree hole, can house many thousands of individuals, all graded within a rigid caste system. Female workers collect food and maintain the colony, while males mate with the queen who lays her eggs in the wax cells of the comb. Communication inside the nest includes an extraordinary 'sun dance', in which workers use a strictly choreographed wiggling dance routine to indicate the precise direction of food in relation to the sun. Though honey is considered a great delicacy by man and beast alike, honeybees can be dangerous when their nest is threatened.

226

FINDING WILDLIFE

A boat trip on the Rufiji River at sunset in Nyerere National Park. (AVZ)

The superb quality of television documentaries can create unrealistic expectations of an African safari. A top-notch one-hour wildlife film might be the product of years spent following a semi-habituated animal, and the high-speed chases and intimate wildlife interaction to which armchair safari-goers have become accustomed are seldom observed so easily in the flesh. It's a gap comparable to watching a sporting highlights package on television and being present at a live match: the former is all action, punctuated with thrilling close-ups and slow-motion replays, but it lacks the immediacy and atmosphere of actually *being* there as events unfold.

Still, where many neophyte safari-goers might expect to encounter wall-to-wall wildlife chomping on the vegetation and on each other, the reality tends to involve long drives that offer nothing more exciting than the odd antelope or squirrel, punctuated by the occasional sudden thrilling encounter with a trumpeting elephant herd or a leopard slinking through a ribbon of riparian forest. In East Africa, there are actually a handful of reserves – notably Ngorongoro Crater, Maasai Mara and parts of the Serengeti – that come close to the 'wildlife around every corner' archetype. More normally, however, spotting all but the most conspicuous of wild animals is a matter of perseverance, skill, experience and luck.

GAME DRIVES

Poor roads, erratic signposting and a widespread reluctance to rent out 4x4 vehicles on a non-chauffeured basis make East Africa poorly suited to self-drive safaris by comparison with southern Africa. Instead, most visitors book themselves onto an all-inclusive guided safari, the cost of which usually includes all transfers, game drives, fuel costs, accommodation and meals – everything, that is, but drinks, tips and what tour operators tend to refer to somewhat obliquely as 'items of a personal nature'.

At the budget and mid-range level, which embraces something like 95% of the East African market, a typical safari will run like a small-scale guided tour: a party

Elephants have right of way when crossing the road. (AVZ)

of anything from two to ten people will be driven along a circular itinerary in a 4x4 vehicle or minibus, accompanied throughout by a driver-guide who conducts all game drives and arranges any other activities included on the itinerary. At the top end of the price scale, however, safaris tend to operate entirely differently, with guests being flown or driven between a succession of small exclusive camps, each of which will normally provide game drives and other activities using highly trained guides with an intimate knowledge of the immediate vicinity.

As a rule, the driver-guides who lead budget (and to a lesser extent mid-range) safaris in East Africa worked as drivers or mechanics before they moved into the lucrative tourist industry. This can be reassuring when your engine starts grumbling in the middle of nowhere, but it also means that a high proportion of driver-guides have little inherent interest in wildlife or any great affinity for the bush. Sure, you can assume your driver-guide will know a lion when they see one, probably a lilac-breasted roller too, but it's unrealistic to expect them to display the passionate sense of vocation or in-depth knowledge associated with the guiding industry in southern Africa. This is seldom a problem, since most driver-guides will do everything they can to ensure a safari runs smoothly, if only in the hope of securing a decent tip. But it does place the onus on you to emphasise any specific interests and dislikes, and to take a proactive role in game spotting, route planning, bird identification, deciding on photographic stops, etc.

TIMING

Typically, a day on safari is structured around a four-five-hour morning game drive and a shorter afternoon drive, returning to camp in between for a leisurely lunch and siesta. A full day drive may be necessary if you want to travel far from camp, and may be unavoidable on days when you travel between different lodges or parks.

Where possible, embark on your morning game drive shortly before sunrise (ideally with a packed breakfast, as provided by most lodges on request) and plan your afternoon game drive so that you return to camp at the latest permitted hour. First and last light are when you are most likely to encounter secretive nocturnal predators such as leopard or serval, while the more conspicuous likes of lion, spotted hyena and jackals are most active (and interactive) before the heat of the day kicks in. Other advantages of being out during the early and late light are the lower volumes of tourist traffic, the cooler temperatures, the higher level of avian activity and the often sumptuous photographic light. Somewhat quirkily, lodges in Ruaha and Nyerere national parks tend to push guests towards taking a leisurely sit-down breakfast prior to the morning game drive – a pressure you'd do well to resist, since dawn is the most atmospheric time of day and offers the best chance of good predator sightings in southern Tanzania as elsewhere.

In national parks, official driving hours typically run from around sunrise to sunset and are normally posted at all park entrances. These hours will be enforced rigidly at busier lodges on the main tourist circuits, but may be treated more laxly at small isolated bush camps. Either way, night drives (or, more accurately, late afternoon drives that continue for an hour or two after sunset) are forbidden in most

Night drives often present the best opportunity for spotting nocturnal species such as the leopard. (AVZ)

East African national parks, but they are sometimes offered in private concessions bordering the official reserves, as well as in privately managed reserves such as those on the Laikipia Plateau.

If you visit a lodge that offers night drives, don't miss out on it. The main attraction is the opportunity to see a host of unusual creatures that are more difficult to locate in daylight – bug-eyed galagos, twitchy elephant shrews, catlike genets and civets, spectral chameleons, the rare striped hyena and aardwolf, roosting nightjars and screeching owls, and – if you're really lucky – the singularly peculiar termite-guzzling aardvark. But even if you see very little in terms of wildlife, the African bush possesses an unforgettable haunted quality after dark, and the sparkling night sky can be utterly mesmerising.

WHERE TO LOOK

Don't leave all the wildlife spotting to your driver-guide. Some guides are excellent at picking out wildlife, others less so, and even the most experienced spotter might overlook a roadside lion if they are focusing on a difficult stretch of road. During the warmer middle part of the day most animals prefer to lie or stand in the shade, so scan the ground below isolated trees in open country, and try to look into thicker bush rather than letting your eyes follow openings through it.

In an open-topped vehicle, you're bound to stay more alert if you stand rather than sit, and the additional elevation will allow you to see further into the bush than a seated driver or passenger. Be conscious of a 'ticking' mentality that informs the way some drivers conduct game drives: if you've seen and photographed one giraffe, for instance, your driver might well decide not to stop for any others, no matter how distant or static that initial sighting was, or how photogenic a subsequent sighting is.

It's always worth stopping for a few minutes at any visible river, reservoir or other watering point. Most animals need to drink at least once daily, so from mid-morning onwards, perennial water bodies tend to attract a steady trickle of thirsty elephant, buffalo, zebra, giraffe and other herbivores, especially during the dry season. Should elephants be hovering at the water's edge, it's worth sticking around to see whether they actually get into the water and start playing, which can be great fun to watch.

Vegetated riverbanks tend to attract fewer transient drinkers than isolated reservoirs, but if you switch off the engine and sit for 10 minutes at an apparently deserted riverbank scene, you'll be surprised at what you see – a crocodile or hippo surfacing, a flash of brilliant colour as a kingfisher or bee-eater swoops from its perch, a bushbuck or kudu emerging from the tangled undergrowth, or a troop of monkeys erupting into treetop activity.

FOLLOWING CLUES

Look for indirect signs of animal activity. An aggregation of circling or roosting vultures will often reveal the whereabouts of a kill, while noisy alarm calls from an arboreal baboon troop or guineafowl flock might well be in response to a prowling predator. In Uganda specifically, male sentries on the periphery of kob herds frequently utter a snorting alarm call to indicate the presence and direction of slumbering lions, while the same inference might be drawn elsewhere from the sight of an impala herd staring in nervous unison.

Bearing in mind that many animals follow manmade roads in the same way that they would normally travel along established wildlife tracks, a trail of fresh paw prints laid on a muddy or sandy road – especially when located in the early morning – will often lead an experienced guide to the predator that made them, while a few steaming piles of roadside dung combined with torn-off branches and other damaged vegetation is a sure sign that elephants passed through recently. Ironically, however, the most conspicuous clue to an interesting sighting in the more popular parks requires no specialist bush knowledge to interpret, consisting as it does of a huddle of 4x4s and/or minibuses further along the road.

The gaze of an alert prey species, such as this kob, often reveals an approaching predator. (AVZ)

TAKING PHOTOGRAPHS

Game drives generally offer the best opportunities to photograph wildlife, because the vehicle doubles as a hide and as a stabilising device for your camera. Some drivers are more skilled at lining up vehicles than others, so it might be worth getting involved in this yourself. In most instances, you'll obtain the best result by approaching along a line that places you more or less directly between the animal and the sun, but that also avoids placing any distracting vegetation between subject and lens. It's worth trying to develop a feel for the right speed of approach: too fast and direct and you might scare off the animal, too slow or stop–start and it might feel like it's being stalked. If you sense the animal is getting twitchy, stop for a minute or two, and if it still doesn't settle, best to leave it in peace.

Most long lenses now have some sort of vibration reduction feature, but if yours doesn't, you can improve its stability (and increase the odds of sharp photos) by using the vehicle as support, ideally in combination with a beanbag. Do be aware, however, that the vibration of a running engine will almost certainly result in a blurred image, as will the shuffling of an antsy-pants driver or anything else that causes the vehicle to

A beanbag placed firmly on a car window frame provides a solid support for even the largest of lenses. (AVZ)

move while the shutter is open. Finally, in an open-top vehicle, avoid the temptation of shooting always through the roof – the elevation may provide a better vantage point for distant animals and perched birds, but where the subject is close to the car and at ground level, you may obtain a better result by shooting it square on from the window.

RESPONSIBLE WILDLIFE WATCHING

Avoid any activity that needlessly disturbs the wildlife or has a negative impact on the environment, and also be prepared to discourage your driver-guide from doing so. Desist from littering, especially from throwing matches or cigarette butts into the bush, which could instigate an uncontrolled fire. Discourage your driver from off-road driving where it is forbidden, and from the moronic practice of hurtling along behind predators while they are on the hunt. Never feed the wildlife: monkeys and baboons in particular will quickly learn to associate humans with food and might then be shot as vermin once they do.

Hooting or yelling at wildlife to gain its attention is another no-no, and allowing your driver to provoke a elephant into mock charge is plain daft. Indeed, when it comes to inflicting artificial noise on the wildlife, their clients and any other vehicles that happen to be around, some driver-guides are quite astonishingly insensitive – unless you intervene, they'll leave the engine running when you stop to look at an animal and let their radio crackle away at full blast, all the while bellowing inanities at their mate in the next vehicle, creating an atmosphere more appropriate to a truck stop than a game reserve.

No thrill quite matches that of encountering the big game on a foot safari. (AVZ)

GUIDED WALKS

In most national parks – other than those in forested or highland areas where 'dangerous game' is absent or uncommon – visitors are confined to their vehicles, except at designated spots such as lodges, rest camps and entrance gates. However, there is no more exciting way of seeing wildlife than on foot, and many national parks, wildlife reserves and private reserves (most notably Nyerere National Park in southern Tanzania) now offer morning game walks led by an armed local guide.

Guided walks offer a far more involving experience than a game drive, often transforming what would otherwise be a relatively mundane sighting (yet another wildebeest or impala) into something altogether more immediate and inspirational – and lending a definite edge to any encounter with elephant, buffalo, rhino, lion and the like. No less important, walking in the bush provides a wealth of stimuli to senses that tend to be muted within the confines of a vehicle – on foot, you are far more conscious of sounds, smells and textures – and offers a chance to concentrate on smaller creatures such as birds, bugs and butterflies, and to examine such environmental minutiae as animal tracks and spider webs.

Animals that are habituated to vehicles are often less relaxed when they encounter human pedestrians,

Forest birdwatching requires binoculars and patience. (AVZ)

so it makes sense to dress in neutral colours such as green, grey or khaki, to take off any conspicuous items of clothing such as brightly coloured caps or scarves, and to desist from dousing yourself in perfume or any other non-functional artificial scents. Lightweight long trousers, socks and solid shoes provide better protection against thorns and biting insects than shorts and open shoes, and a sunhat and sunblock will protect against direct sunlight.

You should also carry drinking water and, depending on the duration of the walk, a snack of some sort (avoiding fruit or meat) and a raincoat and/or sweatshirt just in case the weather turns nasty. As a rule, it is more difficult to approach wildlife closely on foot than in a vehicle, which means you are less likely to make use of a camera and more likely to regret not carrying binoculars. If you have a specific interest, such as birds or butterflies, it's worth asking for a guide with specialist knowledge – or at least making your interest clear before the walk starts.

Even more so than on a game drive, noisy chatterers are likely to scare off any nearby animals and to spoil the excursion for other guests, so try to say as little as possible as softly as possible. As for safety, the risk of being attacked by a wild animal is very small, but it is vital to pay attention during your pre-walk briefing and to listen to your guide at all times – especially in the presence of a potentially dangerous animal.

A walk with a knowledgeable and articulate guide will greatly enhance your understanding of the bush. You'll be shown how to identify the more common and interesting trees, and may have some of their traditional medicinal uses explained to you. Especially in the early morning, the volume and variety of birdsong can be quite overwhelming, and a bush walk offers the opportunity to seek out mixed bird parties comprising various inconspicuous sunbirds, warblers, bush-shrikes and other species.

Butterflies, generally most active from mid-morning onwards, are often abundant near water and forest edges, while large webs made by colourful spiders dangle in the treetops, and creepy-crawlies such as dung beetles and millipedes creep and crawl along the ground. You'll probably want to look out for lizards and tortoises, and while snakes are unlikely to be observed (most would say fortunately), you may well see the odd series of S-shaped ripples created by their undulating method of locomotion on sandy soil.

TRACKS AND SIGNS

Even where large mammals seem to be uncommon, the bush is often littered with evidence of their passing. Droppings, often left in conspicuous places, betray the presence of many. Among the easiest to identify are the gigantic steaming pats deposited by elephant and rhino, the sausage-shaped scats of carnivores such as lion and jackal, the chalky white calcium-rich droppings of bone-chomping hyenas, and the neat piles of tiny pellets that mark the territorial boundaries of a dik-dik or duiker.

Sandy trails or dirt roads are often criss-crossed by an amazing number and variety of animal tracks. Most numerous are the near-symmetrical cloven-hoof marks of antelope, which consist of two teardrop-shaped segments and look rather like an elongated, inverted heart split down the middle. Most antelope prints are too similar to be identified by shape alone, but the possibilities can often be narrowed down when factors such as size, number of prints and environment are taken into account. The print of a duiker or dik-dik is only 2cm long, while that of an eland is up to 14cm, and intermediate species tend to have proportionately sized spoor, with reedbuck, bushbuck and impala prints measuring about 6cm long, and those of wildebeest, oryx and hartebeest around 11–12cm. A solitary trail of medium-sized prints in riparian forest is far more likely to belong to a bushbuck than a reedbuck, while a muddle of trails in light woodland would most likely be made by an impala herd.

Of the other cloven-hoofed ungulates, a buffalo print is larger (around 15cm) and more rounded than that of any antelope, while the giraffe's is squarer, more elongated, and can be anything up to 20cm long. The print of a zebra is very horse-like, consisting of a large horseshoe in front of a pair of small antelope-like teardrops, while a hippo print looks like

Typical carnivore footprints, like these lion's (*top*, WC/LM), consist of a pad with several toes, whereas most large herbivore prints, like this elephant's (*below*, SSp), are more hoof-like in appearance.

a gigantic four-pronged fig leaf. Larger still, rhino prints look a bit like squashed heads with outsized ears, and aren't anything like as big as plain oval elephant prints, which can measure up to 70cm long and are recognisable by size alone.

Look closely on the trail, and you may also pick up the spoor of a carnivore, which typically consists of an inverted heart-shaped pad print and a quartet of oblong or circular toe marks, like the paws of a domestic dog or cat. When trying to identify what made the print, one of the first things to look for is a row of triangular claw marks above the toes – this is lacking in the case of genets and all cats other than the (only semi-retractile clawed) cheetah. As with antelope, the size of the print is generally proportionate to the animal. A clawless, cat-like print of 10cm or longer is almost certainly the spoor of a lion, while a similarly sized print with dog-like claw marks and toes squashed together will have been made by a spotted hyena. The print of a leopard, cheetah or African wild dog is typically 7–9cm long: the absence of claw marks would indicate that a leopard was responsible, while a flurry of overlapping trails would point to the more sociable wild dog.

At the other end of the spectrum, the print of the dwarf mongoose is only 2cm long. Two deceptively proportioned carnivore prints are those of the Cape clawless

Gorilla tracking in western Uganda or Rwanda offers an exciting opportunity to see wildlife on foot. (AVZ)

otter and honey badger, both of which can be up to 8cm long – the latter showing distinctive claw marks well in front of the pad. It would be possible to confuse primate and carnivore prints, but the former almost always resemble a human footprint in general shape and have five clearly defined toes; baboon prints can measure up to 16cm from heel to tip, but most other monkey prints are 4–6cm long.

FOREST WALKS

Although walking is forbidden in most East African savannah reserves, it is permitted and even encouraged in forested reserves, most of which cannot easily be explored by vehicle and harbour lower densities of potentially dangerous large mammals. The gorilla-tracking excursions offered at Bwindi Impenetrable National Park (Uganda) and Volcanoes National Park (Rwanda) are the most renowned of East Africa's forest walks, but chimp tracking is also on offer at perhaps a dozen sites regionally, most prominently Mahale Mountains and Gombe national parks in Tanzania, Nyungwe National Park in Rwanda, and Kibale National Park, Budongo Forest and Kyambura Gorge in Uganda.

Understandably, gorilla- and chimp-tracking excursions are highly goal-orientated and pay little attention to other aspects of the forest fauna and flora. Several national parks and forest reserves, however, also offer more relaxed and low-key forest walks that concentrate on trees, flowers, birds, butterflies and monkeys. Uganda is particularly rewarding in this regard: excellent forest walks led by knowledgeable guides are available at Kibale, Bwindi, Semuliki, Mgahinga and Elgon national parks, as well as at several less protected areas such as Budongo, Mpanga and Mabira Forest. Most of these sites also offer great opportunities for casual unguided walks in forest fringe habitats. Outside of Uganda, good sites for guided and unguided forest walks include Kenya's superb Kakamega Forest, the endemic-rich Amani Nature Reserve in northeastern Tanzania, and the immense Nyungwe National Park in Rwanda.

Hiking is also the best or only way to explore the upper slopes of Kilimanjaro, Kenya, Rwenzori and other large East African mountains, but none of these is primarily (or secondarily) a wildlife-watching destination.

BOAT TRIPS

Wildlife viewing from a boat makes a welcome change from the standard safari regime of two daily game drives, but it is only an option in a limited number of reserves. For 'Big Five'-orientated visitors, the prime aquatic site in East Africa is the Nyerere National Park, where most camps offer two-three-hour excursions on small motorboats, though it is run a close second by the more populist launch trips that run along the Nile below Murchison Falls National Park and the Kazinga Channel in Queen Elizabeth National Park. Large numbers of hippos and crocodile are a certainty in these three places, but there's also a very good chance of elephant, buffalo and various antelope coming down to drink or bathe, and a more remote possibility of seeing big cats at the water's edge.

Other sites that offer game-viewing or birdwatching excursions by motorised boat include Saadani National Park, Rubondo Island National Park, Lake Baringo, Lake Naivasha, Akagera National Park, Lake Mburo National Park and Toro-Semliki Wildlife Reserve, while Arusha National Park is the only major reserve in East Africa where canoe excursions are run by a licensed operator – allowing you to view elephant, buffalo and other wildlife from a boat without the distraction of engine noise. More informally, the Mabamba Swamp near Entebbe and Kampala has become very popular as one of the most reliable sites anywhere for shoebill sightings, and non-motorised boat trips can be arranged easily (and affordably) through local fishermen. You don't need to be an avid twitcher to enjoy exploring this superb wetland.

Safety-wise, your main concern on any boat trip should be the sun. Most boats offer only limited shade, and the intensity of the sun's rays is amplified when they are

The launch trip on the Kazinga Channel is a highlight of any visit to Queen Elizabeth National Park. (AVZ)

Reticulated giraffe *(Giraffa camelopardalis reticulata)* in Meru National Park. (AVZ)

reflected by the water's surface. This means that you are likely to burn twice as quickly as you would on foot or in a vehicle, so wear a hat (ideally one that can be tied around your neck) and douse yourself liberally with sunscreen. Crocodiles pose no real threat unless your boat capsizes, in which unlikely event you should swim directly to the closest shore. A hippo will occasionally attack a canoe, kayak or motorboat, but this isn't a significant risk in a motorboat steered by a local who knows where the hippos usually hang out. There is always a risk of being drenched by a storm or by choppy waters, so carry all valuables and damageable goods in waterproof bags. If you carry a long photographic lens, bear in mind that the constant rocking of a boat makes it too unstable to be a useful support: you'll need to handhold the camera and may have to sacrifice depth of field in order to maximise shutter speed and probable sharpness.

Although the most reliable thrills of a boat trip are close-up sightings of hippos and crocs, it's noticeable that many terrestrial mammals – including elephant and lion – are far more tolerant of a close approach by boat than by car, especially if the engine is cut and the occupants are quiet. Boats also provide a superb platform for viewing waterbirds, from the goliath heron and African pygmy goose to the African fish eagle and pied kingfisher, and a skilful pilot will often be able to get in very close to herons, storks and other large wading species.

Many animals, like these elephants, are less disturbed by passing boats than by land vehicles. (AVZ)

A hide at The Ark in Aberdares National Park offers a superb opportunity to watch African elephants safely at close quarters. (AVZ)

CAMPS AND LODGES

After a long day driving in or between reserves, it is easy (and for many, perhaps desirable) to perceive time in camp or a lodge as downtime – an opportunity to replenish yourself with a quick nap after that early morning start, or to settle down at the poolside bar with a book and chilled drink. But for the dedicated enthusiast, lodge gardens often offer a welcome opportunity to seek out small wildlife on foot. Most camps are home to a range of lizards, frogs, hyraxes, squirrels and small predators, and can be relied upon to provide good birdwatching, especially in the early morning and late afternoon, when fruiting ficus trees and flowering aloe shrubs often attract species that are less easily seen from a fast-moving vehicle.

Many lodges are built alongside excellent natural viewpoints, from where you can scan the surrounding bush for passing wildlife, while others overlook rivers or lakes that attract a steady trickle of transient wildlife. Such aquatic viewpoints are especially worth your time if they are spotlighted after dark, which is also when many camps are visited openly or surreptitiously by scavenging hyenas, bushpigs, genets and others. The lodge-as-hide concept is taken to its logical conclusion at the so-called 'tree hotels' of Kenya's central highlands (Treetops and The Ark), where game drives are forsaken in favour of long afternoons and nights glued to salt licks and small waterholes that attract all the so-called 'Big Five' along with forest specialities such as giant forest hog, bushpig and various birds and monkeys.

Waterhole at Treetops hotel, Aberdares National Park, Kenya. (AVZ)

URBAN HABITATS

At first glance, East Africa's sprawling and chaotic cities don't hold out much promise for viewing wildlife. And it's true enough that you won't see an elephant or lion wandering through the urban heart of Nairobi or Arusha. On closer inspection, however, most African cities can be surprisingly rewarding when it comes to birds and other medium to small vertebrates. Downtown Kampala, for instance, is a prime site for the compellingly ungainly marabou stork, which often nests atop office blocks and perches on street lamps, while nearby Entebbe is home to habituated troops of black-and-white colobus and vervet monkey. Many hotel grounds set in the leafy suburbia of Kampala, Nairobi, Arusha et al are home to colourful turacos, boisterous forest hornbills, bright yellow colonial weavers and elusive chameleons.

Several East African cities also support impressive fruit-bat colonies, while almost every urban wall below an altitude of around 1,500m is likely to host a

Sunbirds, like this male beautiful sunbird (*Cinnyris pulchella*), are among many species that can be lured to gardens. (AVZ)

few bug-eyed, insect-chomping geckos, and the eerie nocturnal call of bushbabies is often heard in suburban Dar es Salaam and Mombasa. More dramatically, the forested Karen Hills fringing Nairobi still support a population of elusive leopards, while Nairobi National Park is famed for its improbable juxtaposition of giraffes, gazelles and even big cats against a heat-hazed backdrop of shimmering skyscrapers. In short, should your itinerary enforce upon you a stray morning or afternoon in urban surrounds, odds are that with a little forethought and initiative, it will present a great opportunity for some low-key wildlife watching.

On Safari

Tourists watching a Maasai giraffe in Tarangire National Park. (AVZ)

Planning your safari is as daunting as you want it to be. At the one extreme, there remain a few tracts of East African wilderness – parts of northern Kenya, for instance, or western Tanzania – that are best tackled in full-blown, multi-vehicular expedition mode. At the less heroic end of the scale, a quick internet search will reveal any number of all-inclusive upmarket safari packages suitable for anybody with the wherewithal to stuff a change of clothes and toothbrush into a daypack, slip a passport into their top pocket, and stumble from a runway into the open passenger door of a waiting Land Cruiser.

When it comes to the above extremes, a healthy bank balance is something of a prerequisite. But somewhere between them lie many less wallet-draining gradations. For those with a well-developed sense of adventure, and more time than money, there is the DIY approach, which involves using public transport and staying at cheap local hotels – though it has to be said that high entrance fees make East Africa's top safari hotspots increasingly inaccessible on a tight budget. The best budget destination (gorilla tracking aside) is Uganda, where park fees are a little lower, there are more opportunities for exploring on foot, and you can get within striking distance of most national parks using public transport.

For less independently minded souls, or for anybody who wants to see wildlife at minimal financial outlay, overland trucks remain a popular option. The large group size on these trucks greatly reduces transportation and fuel costs, and itineraries are designed to avoid unnecessary expenses. The best country for this sort of thing is undoubtedly Kenya, where – in addition to numerous companies offering circular three-to four-week East African safaris out of Nairobi and one-way trips all the way to South Africa – there are still a few budget safari companies that offer affordable week-long truck or minibus safaris taking in the Maasai Mara, Samburu-Buffalo Springs and other key reserves.

A Turkana-bound overland truck makes a pit stop in the deserts of northern Kenya. (AVZ)

A close encounter with a giraffe in Tanzania's Ruaha National Park. (AVZ)

For those whose budget stretches to it, a private safari is a far more pleasant experience. Not only does it allow for a greater degree of autonomy, but it removes the risk of sharing your vehicle with disagreeable strangers. The cheapest private option is a budget camping safari, the price of which generally includes use of a small two-person tent and bedding in a public campsite, services of a vehicle, driver-guide and cook, and all meals, park fees and activities. Some companies arrange semi-luxury or luxury camping safaris, which offer larger tents, more comfortable bedding and more ceremonious meals. In Tanzania specifically, the top camping safaris – generally very pricey – involve the exclusive use of a private 'special campsite', bringing you closer to the bush than any public site.

For unhappy campers, a more popular option is a lodge-based safari. Again, two broad types of accommodation are available: the large 'hotel in the bush' lodges operated by several chains, and more intimate bush camps offering accommodation in standing tents. The former offer a roof over your head at night, restaurant food, and far more comfort than you'd get on any camping safari, but they also tend to have a rather institutionalised ambience that sterilises the surrounding wilderness. The packaged feel and physical solidity of these hotel-style lodges may reassure nervous first-timers, but it undoubtedly diminishes the 'wow factor' of a safari by isolating guests from the immensity of the African night sky and accompanying soundtrack. On the whole, these lodges are aimed at the mid-range safarigoer, and a considerable proportion of their custom derives from organised group safaris.

Bush camps, by contrast, tend to be smaller (anything from 6 to 20 units) and to blend into the surrounding bush in a manner that makes it difficult to forget for a moment that you are on safari. Most bush camps are surprisingly trendy in their style of décor, and many are genuinely luxurious, catering almost entirely to fly-in guests,

An evening on safari might entail a long soak at the luxurious Ngorongoro Crater Lodge (*top*), or watching the stars from a campfire (*above*). (AVZ)

for whom they provide all activities using their own experienced guides. Inevitably, this experience comes with a substantial price tag, but for those who can afford it, there really is no better way of experiencing Africa in the raw. Fortunately for those who can't, there are also a number of more rustic tented camps, particularly in Uganda and northern Tanzania, that combine a bush atmosphere with decent but not luxurious facilities and rates comparable to the chain lodges – the best of both worlds, in the opinion of this writer.

Bearing the above in mind, it's worth researching your options carefully before you actually book a safari. Many safari operators will take the path of least resistance when it comes to planning your itinerary, and will push you towards the lodges they know best or to the more easily booked chain hotels, rather than to smaller lodges that might better suit your requirements.

You should also decide whether you want to book through a local East African ground operator or through an agency in your own country. In most cases, a domestic ground operator will have greater in-depth knowledge of the terrain than an international agent, and can offer better rates because it isn't maintaining costly offices and staff in New York or London or wherever. The advantages of using an international agent are that you can more easily plan your itinerary face-to-face or by phone, you will often get a better flight package than you would independently, and there is less room for vagaries and complications when it comes to payment and insurance.

WHAT TO TAKE
CLOTHING
Clothing will probably constitute the bulk of your luggage, dependent to some extent on your accommodation – you'll need a lot more spare clothing on a ten-day camping safari, for instance, than if you spend a week or two at the sort of small, exclusive lodges where anything you turf into the laundry basket is returned fresh and gleaming within 24 hours. The ideal safari clothing is loose-fitting, lightweight,

informal and made of natural fibres. Bright colours are a definite no-no when tracking wildlife on foot; pale colours are less likely to attract tsetse flies but tend to show dust and dirt more conspicuously.

Many of East Africa's more popular reserves are at a fair altitude, and nights are often cooler than one might expect, so bring a couple of sweatshirts and maybe a windbreaker. Waterproof clothing is a good idea during the rainy season, particularly if you are tracking gorillas or climbing mountains. From dusk onwards, it's advisable to wear closed shoes, socks and long trousers to protect against mosquito bites. Vehicle-based safaris don't generally call for any special footwear, but decent walking shoes or boots are essential for foot safaris, forest walks and mountain ascents. Specialist clothing is essential for serious mountain hikes, eg: Kilimanjaro, Kenya, Meru or Rwenzori.

CAMERA

Few people would consider going on safari without a camera, be it a proper SLR or one built into a phone. Remember, however, that wildlife photography and filming is a specialised field, and the fantastic footage and perfect images to which we're accustomed are generally the product of patience, experience, planning, high-quality equipment and an element of luck. For decent results, an SLR is preferable to a phone, and a high-magnification lens (200mm *at the very least*) is more or less essential. Zoom lenses (eg: 70–300mm) are generally more affordable and allow for greater compositional flexibility than fixed (prime) lenses, but they tend to lack the sharpness of the latter and to lose at least one aperture stop at full magnification, making them less useful in low light conditions. You can increase magnification by using a converter, but with some loss of clarity and a further loss of one/two aperture stops for a 1.4x/2x converter. The bottom line, if you're serious about wildlife photography, is to buy the best long lens you can afford.

Photography is all about positioning. (AVZ)

Obvious as it sounds, do make sure you have all the batteries, plugs, charging cables and storage devices you need, as well as a universal adaptor. A beanbag is unnecessary if you are using a phone or have a lens with good vibration reduction, but otherwise it remains the most flexible and stable option for shooting out of a vehicle. Ideally, buy or make one with a zip so you can fly with it empty and fill it up locally with whatever is available (beans, rice, dried corn). Make sure your camera bag is well insulated against the insidious African dust.

BINOCULARS

A pair of binoculars is essential for viewing distant wildlife and, especially, for watching birds. For dedicated birdwatchers, 8x magnification is the minimal requirement, but 10x, 12x or (only for the steady of hand) 16x is even better. The trade-off between full-size binoculars (eg: 8x40, 10x50) and their compact counterparts (eg: 8x25, 10x30) is that the former have a wider field of vision and tend to show colours more brightly as a result of capturing more light, while the latter are considerably more portable and steady to hold, and they tend to be cheaper. On the whole, you will get what you pay for when it comes to binoculars: cheap or obscure brands often suffer from poor focusing or lens alignment, a distorting or prismatic effect, and dull or inaccurate rendition of colours. If your budget runs to it, it's worth paying a bit more and sticking with a recognised brand. Avoid gimmicky binoculars (with features like zoom or universal focus) at all costs. And remember that binoculars work perfectly well at night, so don't forget to take them on your night drive – indeed, looking at the African night sky through even the most basic binoculars can be a genuine revelation, revealing thousands of stars that are invisible to the naked eye.

OTHER ESSENTIALS

Bring sunblock, water bottle (kept topped up), hat, sunglasses, torch, daypack, a toilet bag with all essentials and a basic medical kit. A penknife or multiple pocket tool may also prove useful. Contact-lens users might find a pair of old-fashioned glasses useful in the dusty conditions. Travel guides and field guides, though available in major cities, are best bought in advance. Mobile-phone users should note that local SIM cards are very cheap, as are local phone calls and data bundles. Be aware that Wi-Fi and other internet connections are still occasionally unavailable in some remote bush destinations, though this is increasingly rare. Ask your tour operator for details.

These days, it is usually possible to pay using a credit card (Mastercard is generally most widely accepted) in urban hotels. Some game lodges also accept credit cards, often depending on how remote they are, but networks can be unreliable in the bush, so it is a good idea to carry some hard currency (ideally US dollars), which you can change to local currency at foreign exchange outlets or use directly where required. Tipping is cash only, for obvious reasons. Etiquette varies greatly: in Tanzania, for instance, safari drivers are paid little so tips are much more important; in Uganda, by contrast, drivers receive a decent salary and tips are more of a bonus. Consult your safari operator in advance.

All wild animals should be treated with caution, including the baboons that often hang around camps, and any protective mother with a baby. (AVZ)

HEALTH AND SAFETY

You should take informed medical advice before planning your trip. Meanwhile, the following basic precautions are worth emphasising here:

Inoculations Check well in advance what jabs or boosters you require.

Malaria This mosquito-borne disease, prevalent throughout East Africa, is the single biggest health threat. It is vital to take prophylaxis, take all reasonable steps to avoid bites (cover up after sunset, use insect repellent and mosquito nets wherever available) and be alert to malarial symptoms after you return home.

Sunstroke and dehydration Wear a hat, use sunscreen and drink plenty of fluids.

Bilharzia Never swim in lakes or rivers without first seeking local advice. The main risk is bilharzia, a nasty disease borne by freshwater snails, but a crocodile attack is also a possibility.

Snakebite Snakes are very shy and bites are rare. You reduce any risk by wearing closed shoes and trousers when out walking, and watching where you put your hands and feet, especially in rocky areas or when gathering firewood. Very few snakebites deliver enough venom to be life-threatening, but it is important to keep the victim calm and inactive, and to seek urgent medical help.

Other bites Many spiders and scorpions deliver nasty bites, but this is largely avoidable by wearing solid shoes (which should be shaken out before they are donned). Such bites aren't usually life-threatening, but extreme cases may require medical attention. Mosquitoes aside, the most annoying biting insects on safari are tsetse flies and horseflies, the latter being harmless to tourists. Check yourself closely for ticks after walking in grassy places – they are usually easy to remove within an hour or two of them clambering on board.

Wild animals Don't confuse habituation with domestication. East Africa's wildlife is genuinely wild: the lions that laze in front of your Land Rover would almost certainly either flee from or attack anybody fool enough to disembark in their presence, and elephant, hippo, rhino and buffalo might all bulldoze a pedestrian if feeling threatened. Such attacks are rare, however, and almost always stem from a combination of poor judgement and poorer luck. A few rules of thumb: never approach wildlife on foot except in the company of a trustworthy guide; never get between a hippo and water; never leave food in the tent where you'll sleep; and be aware that running from a predator can trigger its instinct to give chase.

Car accidents After malaria, dangerous driving is the biggest threat to life and limb in East Africa. On a self-drive safari, drive defensively, being especially wary of stray livestock, pot-holes and imbecilic or bullying overtaking manoeuvres. Many vehicles lack headlights and most local drivers are reluctant headlight-users, so avoid driving at night and pull over in heavy storms. On a chauffeured safari, don't be afraid to tell the driver to slow or calm down if you think they are too fast or reckless.

Crime Opportunistic theft and to a lesser extent mugging provide cause for caution in some East African cities, notably Nairobi (nicknamed 'Nairobbery'), but the biggest risk in game reserves is a room attendant dipping into your wallet to remove a banknote or two. It's advisable to make advance enquiries about any game reserve bordering areas prone to banditry or political unrest.

WHERE TO GO

There are literally hundreds of national parks, game reserves and other protected areas in East Africa, ranging from savannah reserves the size of a small European country to tiny community-protected enclaves of swampland or rainforest, and it would take the best part of a year to explore them all thoroughly. The following accounts provide a country-by-country overview of the region's most popular and ecologically significant reserves, and is designed to direct readers to the country and circuit that will most suit their requirements and interests. For more detailed trip planning, it's advisable to consult a country-specific guidebook or a knowledgeable specialist safari operator.

TANZANIA

Extending over 945,166km^2, Tanzania is the largest country in East Africa, and covers an area greater than Kenya, Uganda and Rwanda combined. It is also one of the most ecologically diverse nations on Earth, and arguably the finest wildlife-viewing country in Africa, boasting three distinct safari circuits, each of which stands as a top-notch destination in its own right. In statistical terms, an unrivalled 25% of Tanzania is accorded formal conservation status, and the country protects roughly 20% of Africa's large mammal biomass, including one of the world's largest elephant populations, and an ungulate migration comprising more than 2 million wildebeest, zebra and gazelle. Furthermore, Tanzania is home to Africa's highest (and fifth-highest) mountain, its most famous national park, its largest game reserve, portions of its three most expansive lakes (all of which rank among the world's top ten freshwater bodies) and the largest intact volcanic caldera on earth. And for those who are unmoved by statistics, the map of Tanzania reads like a litany of Africa's most evocative place names: Zanzibar, Kilimanjaro, Serengeti, Nyerere, Ngorongoro Crater, Olduvai Gorge, Gombe, Lake Victoria, Lake Tanganyika and Lake Malawi among them.

As a safari destination, Tanzania has traditionally been overshadowed by Kenya, a situation that harks back to the colonial era, when it experienced lower levels of European settlement and infrastructural development. In the early post-independence era, Tanzania functioned as little more than a southerly annexe to the Nairobi-based safari industry, prompting President Nyerere's piqued decision to close the border with Kenya in the late 1970s. Short-lived as it was, this closure of ranks backfired in the short term, leading to a massive reduction in tourist arrivals and a gradual decline in the upkeep of game lodges and other tourist facilities. The 1990s, by contrast, witnessed a gradual but steady improvement in every aspect of Tanzania's tourist infrastructure, driven by a rejuvenated industry based not in Nairobi but in the domestic settlements of Arusha, Moshi and Dar es Salaam. This revival has continued into the new millennium, so that Tanzania now vies with Kenya as the second most popular tourist destination in sub-Saharan Africa after South Africa.

The **northern safari circuit** is the linchpin of Tanzania's tourist boom, and comprises the Serengeti, Lake Manyara and Tarangire national parks, Ngorongoro

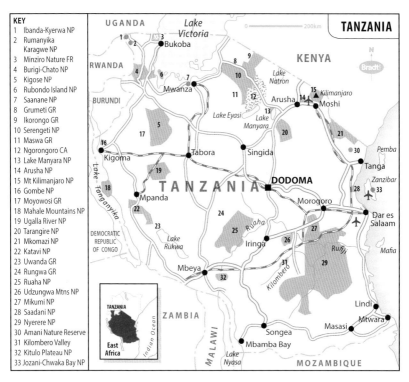

KEY	
1	Ibanda-Kyerwa NP
2	Rumanyika Karagwe NP
3	Minziro Nature FR
4	Burigi-Chato NP
5	Kigose NP
6	Rubondo Island NP
7	Saanane NP
8	Grumeti GR
9	Ikorongo GR
10	Serengeti NP
11	Maswa GR
12	Ngorongoro CA
13	Lake Manyara NP
14	Arusha NP
15	Mt Kilimanjaro NP
16	Gombe NP
17	Moyowosi GR
18	Mahale Mountains NP
19	Ugalla River NP
20	Tarangire NP
21	Mkomazi NP
22	Katavi NP
23	Uwanda GR
24	Rungwa GR
25	Ruaha NP
26	Udzungwa Mtns NP
27	Mikumi NP
28	Saadani NP
29	Nyerere NP
30	Amani Nature Reserve
31	Kilombero Valley
32	Kitulo Plateau NP
33	Jozani-Chwaka Bay NP

Conservation Area, and various lesser known sites. It can be explored over anything from three nights to two weeks, starting from the self-styled 'safari capital' of **Arusha**. Despite its green location at the base of the imposing Mount Meru, Arusha township – with its literally hundreds of safari companies, excellent restaurants and myriad touts and curio sellers – is more notable for its amenities than for any compelling sightseeing. A more alluring proposition for nature lovers is the string of upmarket lodges at nearby **Lake Duluti** and **Usa River**, whose rustic surrounds are enhanced by some great birding and the strong possibility of glimpsing Kilimanjaro's majestic peaks on the northwest horizon.

The small but scenic **Arusha National Park**, only 45 minutes' drive from Arusha itself, can be visited as a day excursion or as an overnight destination in its own right. Highlights of this underrated park include the Momella Lakes and their large concentrations of waterbirds (including flamingos) and the stunning Ngurdoto Crater, a smaller version of Ngorongoro whose jungle-clad slopes usually offer good sightings of blue monkey and black-and-white colobus. Other common wildlife includes giraffe, buffalo, zebra, Kirk's dik-dik and an intermediate population of Defassa and common waterbuck. Capping it all is the magnificently craggy outline of **Mount Meru**, the fifth-highest mountain in Africa, and on a clear day the park also offers great views eastward to Kilimanjaro.

Better views of **Mount Kilimanjaro National Park**'s iconic snow-capped peaks can be had from the town of **Moshi**, 100km east of Arusha. Along with the

Elephants at the Tarangire River, Tarangire National Park (AVZ)

nearby villages of **Marangu** and **Machame**, Moshi is a popular base from which to undertake the five- to six-day ascent of this 5,891m extinct volcano, Africa's highest peak and the world's tallest free-standing mountain. In terms of wildlife, Kilimanjaro is of limited interest, but the vegetation is amazing, the forest zone hosts various monkeys and duikers, while the iridescent scarlet-tufted malachite sunbird (*Nectarinia johnstoni*) inhabits the higher moorland. Even if you don't attempt the ascent, there's no reason why you couldn't stay in Moshi, Marangu or Machame as an alternative to Arusha. True, Kilimanjaro is normally shrouded deep in the clouds, but when these lift – typically in the late afternoon or early morning – it is difficult to imagine a more gaspingly inspirational sight.

A popular first stop on the northern safari circuit, **Tarangire National Park** preserves a classic piece of dry African woodland studded with plentiful baobabs. It comes into its own during the dry season (July to November), when the vast herds of game that roam the Maasai Steppes congregate in the vicinity of the perennial Tarangire River. It's superb elephant country, and the impressive ungulate densities feed a solid population of big cats. A lengthy bird checklist is headed by the endemic ashy starling and yellow-throated lovebird (*Agapornis personatus*), but also includes the red-and-yellow barbets that frequent the park's ubiquitous termite mounds, sometimes alongside chattering bands of dwarf mongoose. Landmarks include the superb clifftop location of Tarangire Safari Lodge – even if you don't stay here, do pop in for a drink – and seasonal Lake Burungi, more often than not a flat, dry expanse of mirage-inducing sand.

Part of the same ecosystem as Tarangire, **Lake Manyara National Park** lies on the Rift Valley floor in a setting extolled by Hemingway as 'the loveliest I had seen in Africa'. Only 330km² in area, it's the archetypal 'grower' – seldom will any given game drive compare to a few hours in the Serengeti, but protracted exposure leaves you with the feeling that here, more than any other Tanzanian reserve, anything could lie around the next corner. The elephant population is among the most substantially tusked and least jittery in Tanzania, and the celebrated tree-climbing lions are observed in arboreal activity with fair regularity. Two-thirds of the park consists of water, at least when the lake is high, but the varied terrestrial habitats embrace lush groundwater forest (olive baboon and the localised blue monkey), open floodplain (large buffalo herds and near-melanistic giraffe) and the rocky escarpment base (klipspringer, particularly around the hot springs at the southern base). And as environmental writer Duncan Butchart has noted: 'If a first-time birdwatcher to Africa [could] visit only a single reserve in Tanzania, then this must surely be it'.

West of Manyara, the **Ngorongoro Conservation Area** protects volcanically formed highlands whose rolling grassy meadows, studded with craggy peaks and forest-enclosed explosion craters, are still grazed by the Maasai and their cattle. The centrepiece is the magnificent **Ngorongoro Crater**, which in its petulant prime stood taller than Kilimanjaro does today. Now it is the world's largest intact caldera, its sheer walls enclosing a 260km^2 expanse of fertile savannah that supports the world's densest populations of lions and spotted hyena, alongside immense herds of wildebeest, zebra, buffalo, gazelle and hartebeest. Massive old elephant bulls, weighed down by tusks of a stature elsewhere sacrificed to the ivory trade, haunt the fever tree groves, while the endangered black rhino is regularly observed in open terrain. Ngorongoro is sometimes dismissed as being 'like a zoo', in reference to the dense vehicle traffic and the high level of mammalian habituation. The latter should be seen as a bonus, since it allows visitors to observe unselfconscious animal

Tree-climbing lions at Lake Manyara, with the escarpment wall behind. (AVZ)

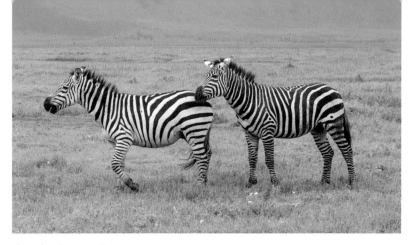

Plains zebras, Ngorongoro Crater. (AVZ)

behaviour at close range, and while the overcrowding can be disturbing, it can also be avoided by descending into the crater at the crack of dawn.

Renowned for the million-strong herds of wildebeest and zebra that undertake an annual migration across its grassy plains, **Serengeti National Park** topped a list of Africa's Top 50 travel experiences published in the magazine *Travel Africa.* Justifiably so. This, quite simply, is Africa's finest reserve, a boundless ocean of koppie-studded grassland that supports incredible densities not only of ungulates but also of predators – indeed, it's not unusual to see lion, leopard, cheetah, serval, spotted hyena, bat-eared fox and a couple of jackal and mongoose species in the same day. The Serengeti is so vast that any sense of overcrowding is restricted to the southeastern plains centred on Seronera Lodge (one of the top three sites in East Africa for leopard, incidentally). More remote from Arusha, the northern and western Serengeti retain a genuine wilderness feel, and visitors with sufficient time and money will gain much from giving these areas at least a couple of nights each.

Several off-the-beaten-track sites could be appended to a northern safari. West of Manyara, **Lake Eyasi** is located in an otherwise arid stretch of the Rift Valley floor inhabited by Tanzania's last hunter-gatherers, the Hadzabe. North of Manyara, the thrillingly scenic **Lake Natron** is a vast primordial sump whose caustic waters provide a breeding ground (inaccessible to humans) to millions of flamingos. Rising above the lakeshore, its steep slopes strewn with white ash in apparent parody of Kilimanjaro, the 2,878m **Ol Doinyo Lengai** (Maasai for 'Mountain of God') is one of the most active volcanoes in Africa, and the ascent to its fascinating caldera, best undertaken at night due to the lack of shade, can be completed in around 6 hours by fit hikers. East of Arusha, **West Kilimanjaro Conservation Area** is a privately managed Maasai concession extending southward from Kenya's Amboseli National Park – an excellent option for high-spenders who want to enjoy game walks and night drives in the elephant-rich shadow of Kilimanjaro. Finally, **Mkomazi National Park**, the seldom-visited Tanzanian extension of Kenya's Tsavo National Park, is now the site of a black rhino sanctuary where you are almost certain to see this endangered creature on a guided visit in one of the park's open 4x4s.

Situated in the southwest of Lake Victoria, peaceful **Rubondo Island National Park** protects the indigenous sitatunga and spotted-necked otter (both easily observed), alongside introduced populations of elephant and giraffe, chimpanzees that have been habituated to tourists, and a varied selection of water and forest birds. In 2019, Rubondo Island was joined by four other new national parks gazetted in the far northwest of Tanzania. Three of these parks, **Burigi-Chato**, **Rumanyika-Karagwe** and **Ibanda-Kyerwa**, lie in remote Kagera District, a wedge of moist and hilly land sandwiched between Lake Victoria and the borders with Rwanda and Uganda. The fourth, **Kigosi National Park**, lies a bit further south. None of these parks is fully developed for tourism as yet, but Burigi-Chato (now the country's fifth largest national park) in particular is a worthwhile off-the-beaten-goal for overlanders driving between Tanzania and Uganda or Rwanda. This part of the country is also home to **Minziro Nature Forest Reserve**, which supports around 50 Guinea–Congo biome bird species unrecorded elsewhere in Tanzania.

Tanzania's so-called **southern safari circuit** is a more disjointed entity than its northern counterpart, and is most easily explored by flying between the various small camps, which provide their own guided game drives in the style typical of southern Africa rather than northern Tanzania. The volume of game in the southern reserves tends to be lower than in the north, but this is compensated for by the relatively untrammelled feel, and the wide variety of activities (fly-camping, guided game walks, boat trips or night drives) that are largely forbidden in the more populist north.

The linchpin of the southern circuit, **Nyerere National Park** (formerly Selous Game Reserve) is one of the largest of its kind in Africa, a 30,893km^2 tract of wilderness set at the core of a thrice-larger ecosystem supporting a now much reduced population of 15,000–20,000 elephants. As an ordinary tourist, however, the much publicised fact of Nyerere's size can feel a little overstated, since the park's dozen upmarket tourist camps are clustered within 5% of its total area! Still, if Nyerere is not quite as exclusive as the hype suggests, it does boast several superb assets. There is the Rufiji, one of Africa's most mesmerising waterways: sandbanks lined with outsized crocs, palm-fringed banks massed with thirsty herds of elephant and buffalo, sluggish water teeming with grunting hippos, and a veritable showcase for Africa's rich aquatic avifauna. Back on dry land, more than 10% of the free-ranging African wild dog population is centred on Nyerere, and diurnal lion kills are unusually commonplace. The range of activities is also excellent: motorboat trips offer a thrilling hippo's-eye perspective on the great river, while foot safaris, led by armed rangers, routinely involve close encounters of the pachydermal kind. The unprotected **Kilombero Floodplain** immediately west of Nyerere is home to at least 40% of the continent's puku antelope, and three endemic bird species. An ecotourism project offers organised trips to this remote area, and backpackers have limited access from the Ifakara ferry.

Ask a seasoned East African safarigoer which is their favourite game reserve, and the answer will probably be **Ruaha National Park**, the country's second-largest national park, whose wild and untrammelled mood is embodied by the bulbous baobab silhouettes that haunt its semi-arid plains and rocky slopes. Ruaha is one of the few reserves where visitors might encounter all three large cat species, and it also supports vast herds of elephant, though these tend to be skittish due to poaching. Transitional to the southern miombo and northern savannah biomes, Ruaha

Wildebeest migration, Serengeti National Park. (AVZ)

harbours an impressive tally of antelope species: Grant's gazelle and lesser kudu at the south of their ranges, and greater numbers of sable, roan and greater kudu than occur anywhere further north. The handful of small lodges in this park are set miles apart, enhancing the wilderness mood and making it the ideal fly-in companion to Nyerere.

Definitely worth a night or two if you are travelling between Dar es Salaam and Ruaha by road, **Mikumi National Park**, transected by the main surfaced road through southern Tanzania, is effectively a westerly extension of Nyerere National Park, and game viewing can be excellent on its northern plains, which host substantial elephant and lion populations, along with large herds of grazers. Mikumi's game lodges are among the cheapest and most accessible in Tanzania. The remote **western safari circuit**, set in the Lake Tanganyika hinterland, consists of three national parks, two of which are renowned for the opportunity to track habituated chimps in their forested home. The tiny **Gombe National Park**, site of the pioneering behavioural research programme established by Jane Goodall in the 1960s, can easily be reached from the port of Kigoma on public transport or private motorboat. Far larger than Gombe, the spectacularly scenic and isolated **Mahale Mountains National Park** runs from the sandy lakeshore to forested peaks that tower almost 2,000m above it. It harbours the greatest variety of primates of any Tanzanian national park, including an estimated 1,000 chimps.

Fly-in trips to Mahale are ideally combined with a stay at the 4,500km² **Katavi National Park**, arguably East Africa's best-kept game-viewing secret, especially between April and November. This is when its seasonal swamps subside to uncover wide grassy floodplains meandered through by feeble but life-sustaining streams where 200-strong hippo pods jostle for wallowing space in any pool that's sufficiently deep to wet a knee in. Lion and elephant sightings are excellent, thousand-strong herds of buffalo regularly amass on the plains – yet until recently you could explore

Katavi for days without encountering another vehicle. This park is starting to catch on, but it remains an exclusive and largely untrammelled wilderness experience, albeit not cheap to visit.

The superb **Indian Ocean coastline** of Tanzania is best known for its idyllic palm-lined beaches, and the ancient Swahili settlements and ruins at Kilwa, Bagamoyo, Pangani, Tanga and of course Zanzibar. More developed parts of the coast also offer some superb diving and snorkelling opportunities, and the possibility of seeking out other marine wildlife that falls beyond the scope of this safari guide, including dolphins and turtles. Opportunities for viewing terrestrial wildlife are more limited, one noteworthy exception being the readily accessible and well-organised **Jozani-Chwaka Bay National Park**, which forms the main stronghold for Zanzibar's endemic Kirk's red colobus monkey.

Offshore from Zanzibar, **Chumbe Island Sanctuary** is the most easterly habitat of the world's largest terrestrial invertebrate, the tree-climbing coconut crab (*Birgus latro*), as well as supporting a small introduced population of the highly endangered Aders's duiker, an East African coastal endemic. For a more conventional coastal safari experience, the **Saadani National Park** is the one place in East Africa where you stand a (slim) chance of seeing elephants or lions walking along the beach, while boat trips through mangrove-lined creeks offer good birding and hippo sightings, and game drives offer your best chance in northern Tanzania of seeing the secretive red duiker and magnificent sable antelope.

Ecologically, the most notable feature of eastern Tanzania is the Eastern Arc Mountains, whose high level of endemicity is not reflected in the inadequate degree of protection they are accorded. A notable exception is the under-publicised **Amani Nature Reserve**, which lies in the Eastern Usambara Mountains inland

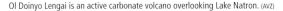

Ol Doinyo Lengai is an active carbonate volcano overlooking Lake Natron. (AVZ)

left Chimpanzee, Mahale Mountains National Park (AVZ)

below Nyerere and Ruaha are East Africa's last strongholds of the endangered African wild dog. (AVZ)

of Tanga, and protects some of the most important montane forest in Tanzania, home to a wealth of rare and endemic birds, mammals, butterflies and other creatures. Accessible by public transport and serviced by knowledgeable local guides and several walking trails, this is a particularly alluring spot for birders, with 340 (including 12 globally threatened) species recorded and the likes of green-headed oriole, Amani sunbird (*Hedydipna pallidigaster*) and long-billed tailorbird (*Artisornis moreaui*) present in the resthouse gardens.

Further south, the **Udzungwa Mountains National Park** was gazetted in 1992 to protect part of the Udzungwa Mountains and its host of endangered endemics, which include three primate and two bird species as well as numerous plants and invertebrates. Good hiking opportunities, easy access by public transport and affordable accommodation at the entrance gate add up to a good destination for backpackers, but several 'Udzungwa specials' such as the endemic hill partridge are actually more likely to be seen in the less accessible **Ndulundu Forest Reserve**, which may be annexed to the national park following the discovery of the kipunji monkey there in 2003.

Part of the kipunji's range has also been incorporated into southern Tanzania's 135km² **Kitulo Plateau National Park**, which became the first such entity in tropical Africa to be set aside primarily for its floral significance when it was gazetted in 2006. Situated to the southeast of Mbeya and known locally as *Bustani ya Mungu* (God's Garden), the montane grassland over Kitulo hosts superb wildflower displays between November and April, and is also of avian significance as a breeding site for the endangered blue swallow (*Hirundo atrocaerulea*) and Denham's bustard (*Neotis denhami*).

Above The giant coconut crab, the world's largest terrestrial crustacean, is easily seen on Chumbe Island offshore from Zanzibar. (AVZ)
Below Sanje Waterfall drops more than 150m. (AVZ)

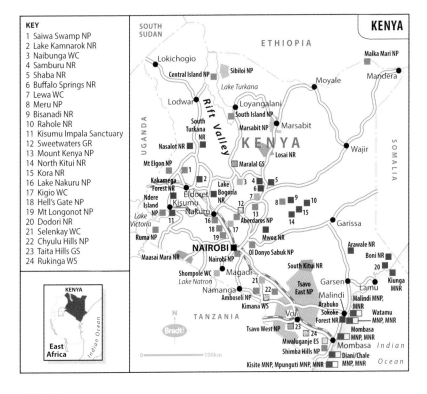

KEY
1 Saiwa Swamp NP
2 Lake Kamnarok NR
3 Naibunga WC
4 Samburu NR
5 Shaba NR
6 Buffalo Springs NR
7 Lewa WC
8 Meru NP
9 Bisanadi NR
10 Rahole NR
11 Kisumu Impala Sanctuary
12 Sweetwaters GR
13 Mount Kenya NP
14 North Kitui NR
15 Kora NR
16 Lake Nakuru NP
17 Kigio WC
18 Hell's Gate NP
19 Mt Longonot NP
20 Dodori NR
21 Selenkay WC
22 Chyulu Hills NP
23 Taita Hills GS
24 Rukinga WS

KENYA

To many outsiders, Kenya is synonymous with East Africa. Never mind that its legendary Maasai Mara is actually a relatively small northern annex to Tanzania's Serengeti, or that Kilimanjaro's snow-capped peak lies on the other side of the same international border, Kenya is the original home of the safari, as immortalised in the likes of *Out of Africa* and *White Mischief*, and it remains in many minds *the* place to see African wildlife. By and large, this reputation is richly deserved: Kenya is a compelling travel destination, boasting a compact but varied selection of tourist attractions that can be explored as luxuriously or adventurously as you choose. Most-first time visitors centre their attention on a south-central circuit combining the **Maasai Mara** and Samburu wildlife reserves with one of more of the Rift Valley lakes, and possibly an overnight stop at one of the 'tree hotels' in the central highlands but the country also boasts an excellent eastern safari circuit, centred on Tsavo and Amboseli national parks, and the relatively untrammelled north and west also offer rich pickings for wildlife enthusiasts.

Effectively a northern extension of Tanzania's vast Serengeti ecosystem, the Maasai Mara is Kenya's most popular game reserve and the one that the purists most love to hate. Fair enough, the Mara (as it's widely known) does carry a heavy tourist traffic compared with most of its counterparts, but for the very good reason that no other East African reserve protects comparable game densities. The best time to

Blue wildebeest at dusk on the Maasai Mara (AVZ)

visit is between July and October, when the wildebeest migration crosses over from Tanzania, offering visitors unforgettable sightings of these manically braying, croc-dodging beasts as they cross the Mara River. But the reserve supports a profusion of wildlife throughout the year, including large numbers of buffalo, zebra and various antelope, as well as a solid population of elephant and a few relict black rhinos. Above all, the Mara is a fantastic reserve for predators, in particular cheetah, spotted hyena and prides of 20-plus lions, whose high degree of habituation makes them easy to watch and photograph.

Lake Nakuru National Park protects the most famous of the string of lakes that lines the Rift Valley west of Nairobi, mainly thanks to its concentrations of up to 2 million flamingos, an avian spectacle to match the wildebeest migration of the Mara. Encircled by hills and fringed by yellow fever trees, Nakuru is a truly gorgeous spot, especially when viewed from the surrounding cliffs, from where the individual flamingos merge into a solid shimmering pink band separating the alkaline water from its bleached rim. Fenced in its entirety, this park is an important relocation site for endangered animals, and a good place to see both species of rhino and Rothschild's giraffe.

The flamingos associated with Nakuru have maintained an erratic presence since the 1990s, and larger numbers are often amassed at the nearby **Lake Bogoria Wildlife Reserve**, which is also one of the few Kenyan strongholds for the magnificent greater kudu. Between Nairobi and Nakuru, the stunningly beautiful and unusually budget-friendly **Lake Naivasha**, though not officially protected, is renowned for its dense hippo population and exceptionally varied birdlife, and a good base for a day hike to **Hell's Gate National Park**, where a variety of large mammals can be seen on foot. To the north of Nakuru, **Lake Baringo** supports prodigious hippos and crocs, and its arid surrounds offer excellent dry-country birding.

North of the Rift Valley, the **Aberdares** and **Mount Kenya national parks** lie in lush moist highlands

A lesser flamingo in Kenya's Lake Nakuru National Park. (AVZ)

Elephants walk past Mountain Lodge on the forested outskirts of Mount Kenya National Park. (AVZ)

dominated by the ragged glacial peaks of Africa's second-highest mountain. The main attraction of these parks in terms of wildlife viewing is a pair of 'tree hotels' – Treetops and The Ark – which overlook floodlit waterholes attracting a steady stream of wildlife, particularly after dark. All the Big Five are present, and elephant, buffalo and black rhino regularly visit the waterholes at night. Lion and leopard are more occasional. Other wildlife includes blue monkey, black-and-white colobus, giant forest hog, bushpig, genet, white-tailed mongoose, waterbuck, the seldom seen mountain bongo, and forest birds such as Hartlaub's turaco. Another popular destination in this area, **Laikipia Plateau** consists of more than a dozen more-or-less contiguous private conservancies stretching over some 9,000km² northward from Mount Kenya into the semi-arid savannah bounding the Ewaso Ng'iro River. Geared primarily towards well-heeled safari-goers wanting to escape the tourist treadmill, the small lodges typical of Laikipia run their own guided game drives in open vehicles, and also usually offer activities such as night drives, horseback excursions and camel rides. Home to an estimated 25% of the world's Grevy's zebra population, Laikipia also provides an important refuge for rarities such as African wild dog and Jackson's hartebeest, and all the so-called Big Five are present.

North of Mount Kenya, the **Samburu-Buffalo Springs** complex of reserves lies in a region of austere, rock-strewn scrubland through which the forest-fringed Ewaso Ng'iro River forms a bizarrely luxuriant ribbon. A host of dry-country 'specials' ensure Samburu boasts the most unusual fauna of East Africa's major savannah reserves, with the likes of Grevy's zebra, reticulated giraffe, lesser kudu, Beisa oryx, Guenther's dik-dik and gerenuk all quite common, while vulturine guineafowl, Egyptian vulture and golden-breasted starling top a long list of localised dry-country birds. Samburu supports healthy populations of elephant and lion, and it must rank with the top handful of East African reserves for leopard sightings. Adding a splash of colour to this otherwise harsh landscape, the Maasai-affiliated pastoralists who share their name with the reserve are often to be seen herding their cattle in the area, and several nearby traditional villages welcome tourist visits.

Protecting a similar mammalian fauna to Samburu but in a more lush and conventionally beautiful landscape, **Meru National Park** lies to the east of Mount Kenya and is cut through by several streams that rise on its slopes. Home for years to George and Joy Adamson, Meru suffered heavily in the poaching wars of the late 1980s and was practically off-limits to tourists for most of the 1990s. Now safe to visit, Meru has been restored to something approaching its former glory by an extensive translocation programme, and it remains the best place to see reticulated giraffe and the skittish lesser kudu. Its poacher-proof 84km² rhino sanctuary is home to introduced rhinos of both species.

The other main victim of the 1980s poaching wars was the 22,000km² **Tsavo National Park**, whose eastern and western components are separated by the main road between Nairobi and Mombasa. In the space of a decade, the park's 40,000-strong elephant population was reduced to just 4,000, while its 6,000 black rhino were poached close to local extinction. Tsavo's resurrection since those gloomy days is little short of miraculous: the elephant population now stands at around 12,000 and 100 rhinos have been reintroduced – some free-ranging, the rest coddled in a large stockade. Landmarks include the game-rich Aruba Dam and bald Mudanda Rock in Tsavo East, and the Sheteni Lava Flow and unique underwater viewing tank at the hippo-populated Mzima Springs in Tsavo West. Tsavo is famous for its red elephants, which acquire their unusual coloration from the red dust. Tsavo East is the more interesting reserve for dry-country 'specials', but Tsavo West offers the better conventional game viewing, complemented by superb views across to Kilimanjaro on the Tanzanian border.

Maasai giraffes in Tsavo East National Park. (AVZ)

Amboseli is home to some very big tuskers. (AVZ)

The ultimate location for Kilimanjaro views, **Amboseli National Park** is also famed for its unusually relaxed and heavily tusked elephants, which were largely unaffected by the ivory poaching of the 1980s. Somewhat desolate in the harsh midday light, Amboseli can be spectacular at dusk, when a suspension of fine volcanic dust (deposited by Kilimanjaro in its fiery pomp) refracts the dying sunlight into a festival of orange and red hues. It's at dusk, too, that the cloudy shroud hovering over the park's southern horizon most often dissipates to reveal Kilimanjaro's snow-capped dome at staggering close proximity. Elephants aside, Amboseli supports abundant wildebeest, zebra, gazelle and giraffe, while spotted hyena, cheetah and to a lesser extent lion are the most visible predators. A memorable feature of this park is its marshes, which contrast strongly with the dusty plains, and support a wide array of waterbirds.

Kenya's coast is renowned for its beaches and reefs, but it also boasts one underrated game-viewing gem in the form of **Shimba Hills National Reserve**, less than an hour's drive from the seaside bustle of Diani Beach. Predators are thin on the ground here, but elephants still roam the lush green hills, along with Kenya's only population of the handsome sable antelope. A superb Treetops-style lodge provides a playground for red coastal squirrels (*Paraxerus palliatus*) by day and bushbabies by night, while a tantalising list of coastal forest birds is headed by the green-headed oriole.

Northward towards Malindi, the **Arabuko-Sokoke Forest Reserve** is home to two endemic birds, Clarke's weaver (*Ploceus golandi*) and Sokoke scops owl, and may still harbour a small population of Aders's duiker (otherwise known only from Zanzibar). Set within this coastal forest, **Gedi Ruins** is the best place to seek out the endangered golden-rumped sengi, a fantastic-looking creature that's occasionally seen bouncing daftly along the forest paths. Further north, and altogether more remote, **Tana River Primate Reserve** provided the sole refuge to the endemic Tana River red colobus and Tana River mangabey, prior to being degazetted in 2007. Global populations of both these monkey species are estimated at fewer than 1,000.

Western Kenya sees few tourists but it does boast two highly worthwhile and backpacker-friendly sites in the form of the **Kakamega National Reserve** and **Saiwa Swamp National Park**. Kakamega offers great primate viewing, with abundant black-and-white colobus and, using a spotlight, a good chance of picking out the nocturnal potto. It is also arguably Kenya's most alluring destination for butterfly and bird enthusiasts, with a checklist of 300 forest birds including some 30 species at the very eastern extent of their range. Saiwa Swamp, Kenya's smallest national park, forms an obvious extension to a trip to Kakamega. Overlooked by wooden viewing platforms and enclosed by riparian forest, this is one of the best places in Africa to see the localised sitatunga and De Brazza's monkey. Kenya's arid north remains one of Africa's most profoundly traditionalist regions, inhabited by a miscellany of nomadic pastoralists such as the Samburu, Turkana, Gabbra and Borena. In the midst of this aridity, an isolated forested massif called **Marsabit Mountain** is protected in an eponymous national park and serviced by what must surely be East Africa's quietest game lodge, with a spectacular location above a forest-fringed crater lake teeming with birds and visited daily by elephants. Turn left at Marsabit, keep driving until there are no trees within sight, and you're in the **Chalbi Desert**, a vast flat nothingness crossed by the occasional skidding ostrich or oryx, but otherwise bereft of visible life except on the rare rainy occasion when it is transformed into a shallow seasonal lake. This western rim of this land of mirages and salt flats drops to the Rift Valley floor and the infinitely mysterious **Lake Turkana**, the world's largest desert lake, whose stunning deep-green waters are hemmed in by an apocalyptic moonscape of extinct volcanoes and naked lava flows. Turkana is of greater interest culturally than for wildlife, though it reputedly supports the world's densest crocodile population (and some of the largest individual crocs on record), and the ornithologically minded will enjoy its disorientating mix of dry-country and water birds.

Turkana huts at Loiyangalani, the largest village on the shore of Lake Turkana. (AVZ)

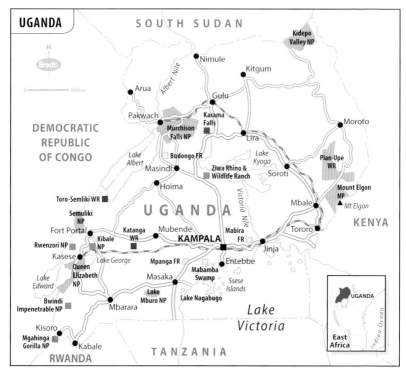

UGANDA

Boasting a lush equatorial setting on the elevated plateau that divides the two main arms of the Great Rift Valley, Uganda is where the eastern savannah meets the western rainforest. True, this comparatively small African country (235,796km²) isn't a classic 'bush' destination in the mould of Kenya or Tanzania, but the well-watered mosaic of rainforest, savannah, montane and wetland habitats makes it a genuine contender for the accolade of Africa's most varied wildlife-viewing destination. Better still, its intimate attractions are unsuited to mass tourism, and thus remain largely unaffected by it. If, as a Birdlife International publication notes, avian variety forms 'one of the most reliable indicators of biological richness', then Uganda's status as by far the smallest of the quartet of African countries where more than 1,000 bird species have been recorded speaks for itself. So, for that matter, does a butterfly checklist that tops the 1,000 mark – probably the largest for any African country, and 20 times more species than have been recorded on the comparably sized British Isles. And there's more to Uganda than bugs and birds: where else, over the space of a few days, could one hike to within a few metres of mountain gorillas and chimpanzees, navigate tropical waterways bustling with hippo and elephant, explore open plains grazed by thousands of antelope and buffalo, and pay respect to Africa's most enthusiastic tree-climbing lions?

All international flights land at **Entebbe**, the most organic of former African capitals, its blink-and-you'll-miss-it town centre flanked by a sprawling golf course and lush botanical garden carved out of the equatorial jungle fringing Lake Victoria. For wildlife lovers, Entebbe makes for a preferable base to the nearby capital city of Kampala: troops of black-and-white colobus and vervet monkey swing acrobatically through relict forest patches where flocks of great blue turaco babble away in the canopy, and fish eagles stand sentinel on exposed branches, occasionally emitting their ringing duet. Entebbe's inherent appeal is boosted by its proximity to the **Ngamba Island Chimpanzee Sanctuary**, a somewhat zoo-like set-up founded by the Jane Goodall Institute in 1998 as a refuge for chimpanzees confiscated from poachers. And only an hour's drive away, a dugout trip into the expansive **Mabamba Swamp** comes with the best odds in Africa of encountering the shoebill, alongside the likes of African pygmy goose, goliath heron and blue-breasted bee-eater (*Merops variegatus*).

Heading southwest of Entebbe/Kampala, the main asphalt road towards the towns of Mbarara and Kabale flanks two minor sanctuaries. First up, near the small town of Mpigi, **Mpanga Forest Reserve** is an attractive spot for a couple of nights' inexpensive camping, and a good place to seek out the pretty red-tailed monkey alongside forest birds such as African grey parrot, blue-breasted kingfisher (*Halcyon malimbica*) and various vociferous hornbills.

Chimps are present in about a dozen Ugandan parks and reserves. (AVZ)

Once poached close to extinction, lions are now a common sight on the plains of Queen Elizabeth National Park, set below the mighty Rwenzori Mountains. (AVZ)

Ideally placed for an overnight stop between Kampala and Bwindi, **Lake Mburo National Park** is often punted as the last Ugandan stronghold for the impala (a distinction that's unlikely to have any southern African visitor salivating in anticipation), but it also harbours prolific numbers of buffalo, giraffe, warthog, eland, zebra and impala, and the boat trips on the lake are a treat. The only acacia-dominated reserve in Uganda, Mburo hosts several bird species found nowhere else in the country.

Situated in the southwestern highlands a few kilometres outside Kabale, **Lake Bunyonyi**, a 30km-long blocked river system, whose irregular shoreline follows the contours of the steep hills that enclose it, is better known as a chill-out spot than for scintillating wildlife viewing, but it does support a varied avifauna, and the normally elusive spotted-necked otter is remarkably common.

The highlight of most visits to southwestern Uganda is tracking mountain gorillas in **Bwindi Impenetrable National Park**, which protects approximately 45% of the global population of this endangered creature. There is more to Bwindi than gorillas, however: the wide tracks around the park headquarters offer superb forest birding, with several of the park's 23 Albertine Rift endemics likely to be identified in the company of a knowledgeable local guide, along with dozens of butterfly species and localised forest mammals such as the outsized yellow-backed duiker and handsome L'Hoest's monkey. Deeper into the park, Mubwindi Swamp is the only non-Congolese locality for the eagerly sought Grauer's broadbill, and it is also the favoured haunt of the park's small population of forest elephant.

Protecting the Uganda portion of the Virunga Volcanoes, **Mgahinga Gorilla National Park** also offers gorilla-tracking, but less reliably so than Bwindi, given that the habituated troop here often crosses into Rwanda for months on end. Other attractions of this magnificently scenic montane park are guided hikes to the Virunga peaks and forest walks in search of the endemic golden monkey.

North of Bwindi, **Queen Elizabeth National Park** is arguably the single most ecologically diverse reserve in Africa – using the avian test alluded to above, you're looking at a checklist of 600-plus species, exceeding that of many African reserves ten times its size (Kruger, Hwange, Serengeti and Nyerere, to name a few). You'd need a full week to explore this fabulous park in its entirety, but popular highlights include the launch trip down the hippo-infested Kazinga Channel, chimpanzee tracking in the forested Kyambura Gorge or nearby Kalinzu Forest Reserve, looking for tree-climbing lions on the southerly Ishasha Plains, and the lush crater lakes and million-strong bat colony in the Maramagambo Forest. The compact network of game-viewing tracks that runs above the Kazinga Channel is an increasingly reliable spot for leopard, and one of the very few places where giant forest hog are regularly sighted by day. Mweya Lodge, set on a bulbous peninsula flanked by Lake Edward and the Kazinga Channel, is surely one of the most sumptuous spots in East Africa, no less so when large herds of buffalo and elephant congregate on the opposite shore of the channel for their afternoon refreshment.

On a rare cloudless day, Mweya also offers distant views of the glacial peaks of Ptolemy's Lunae Montes (Mountains of the Moon), three of which rise to above 5,000m. Now protected within the **Rwenzori Mountains National Park**, this 120km-long mountain range is a popular but challenging destination with dedicated hikers and climbers. The wildlife viewing is limited, as with all Africa's high mountains, but other attractions include several endemic plants and small vertebrates in its lush forests, and the appropriately lunar-looking landscapes of the Afro-alpine zone.

Situated close to Fort Portal, **Kibale National Park** is the most accessible of Uganda's major forests on public transport, and it offers the best chimpanzee tracking in the country, with a better than 90% chance of spotting these charismatic humanlike apes on any given day. The main road through the forest is an excellent place to

A crater lake in the Rwenzori foothills. (AVZ)

At Murchison Falls the Nile is funnelled through a narrow cleft in the rocks. (AVZ)

look for other diurnal primates, including grey-cheeked mangabey, red colobus and black-and-white colobus, and it's not unusual to see 20 butterfly species fluttering around one puddle. A guided walk through the adjacent Bigodi Wetland Sanctuary, an exemplary community-run project, provides a great introduction to Uganda's forest fauna, with up to 50 forest and swamp birds likely to be seen on a good day, including papyrus gonolek and several colourful turacos and barbets.

Easily visited in conjunction with Kibale National Park, the field of 30-odd crater lakes scattered around the small town of **Kasenda** is a scenic spot that offers much to budget-conscious hikers, while relict forest patches such as Lake Nkuruba Nature Sanctuary (another estimable community-run project) harbour black-and-white colobus and a variety of forest birds. A somewhat more esoteric goal from Fort Portal is the Semliki Valley, where the **Semuliki National Park** protects an extension of the Congolese Ituri Forest and 50-odd bird species found nowhere else in East Africa, while the moist savannah of the **Toro-Semliki Wildlife Reserve** lies at the base of a stunningly scenic stretch of Rift escarpment sandwiched between the northern Rwenzori Foothills and Lake Albert.

At the northern end of Lake Albert, **Murchison Falls National Park**, Uganda's largest conservation area, is named after the spectacular waterfall declared by Sir Samuel Baker to be 'the most important object through the entire course of the While Nile'. Launch trips along the lushly vegetated river to the base of the falls seldom disappoint: a preponderance of hippos, gape-mouthed crocs and waterbirds can be guaranteed, elephant, buffalo and various antelope often come down to drink, and black-and-white colobus are resident in the riparian woodland. North of the Nile, a circuit of tracks leads through rolling borassus grassland to lakeshore plains inhabited by plentiful elephant, buffalo, Uganda kob, Defassa waterbuck, Jackson's hartebeest and oribi. Lions are regularly observed, and it's the only place in East Africa where you're likely to see herds of 50-plus giraffe, as well as troops of the spindly savannah-dwelling patas monkey. The tracks overlook the Nile Delta where a few resident pairs of shoebill are sometimes seen from the land, but more reliably observed by boat.

The **Budongo Forest Reserve**, which abuts the park's southern border and is transected by the main road to Kampala, is another superb forest birding site and also offers affordable chimpanzee-tracking excursions. White rhino, extinct in the wild in Uganda, can be tracked at **Ziwa Rhino and Wildlife Ranch**, which lies on the Kampala road south of Budongo and Masindi.

The eastern half of Uganda is of less interest for wildlife, but there are a few scattered highlights. Bisected by the main Kampala–Jinja road, the accessible and affordable **Mabira Forest Reserve** offers good monkey viewing and superb birding, with a fair chance of glimpsing the diminutive blue duiker. The lush stretch of the Nile immediately downstream of Jinja, is rapidly emerging as East Africa's 'adrenalin capital' (activities include rafting grade 5 rapids, kayaking, bungee jumping, quad biking, etc) but the area supports little wildlife other than birds.

Situated on the Kenyan border overlooking the town of Mbale, **Mount Elgon National Park** is an underrated hiking destination, and sites such as the **Sipi Falls** and **Kapkwai** in the foothills are popular with travellers who don't want to hike overnight, with the **Forest Exploration Centre** at the latter being a particularly good spot for forest birds and primates. The little visited **Pain-Upe Wildlife Reserve** north of Mount Elgon is good for dry-country birds and antelope, and it harbours the localised patas monkey, but facilities are limited to the one public road that bisects it. **Kidepo Valley National Park** on the Sudanese border, the most remote conservation area in Uganda, is now serviced by two upmarket lodges and a moderately priced camp. It protects a scenic area of semi-arid grassland inhabited by large herds of elephant and buffalo, decent numbers of lion and leopard, many creatures found nowhere else in Uganda – ranging from cheetah, black-backed jackal and bat-eared fox to greater and lesser kudu – and numerous dry-country birds.

White rhinos in Ziwa Rhino and Wildlife Ranch. (AVZ)

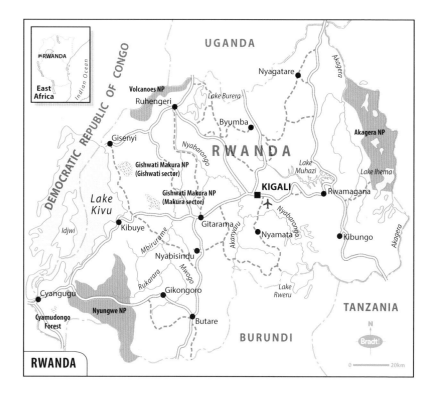

RWANDA

RWANDA

The most densely populated country in Africa, with 12 million people squeezed into its 26,340km² (less than half the size of Scotland), Rwanda is infamous as the site of the 1994 genocide that claimed an estimated million lives. Since the present government assumed power, however, it has staged a miraculous social and economic recovery, and now enjoys a level of political stability comparable to its easterly neighbours. The terrors of 1994, while certainly not forgotten (witness the many chilling genocide memorials scattered countrywide), are gradually receding into history so far as outsiders need be concerned.

As might be expected of such a small and densely populated country, Rwanda is not exactly over-endowed with unspoilt natural resources, but two of its three national parks rank with the very best East Africa has to offer, and it remains arguably the world's premier gorilla-tracking destination. In addition, the stunning vistas that earned it the colonial soubriquet 'Land of a Thousand Hills' are still largely intact, and freshwater bodies such as the Kivu and more serpentine Lake Burera are well worth exploring.

Rwanda's most popular tourist attraction is **Volcanoes National Park**, the site of Dian Fossey's celebrated gorilla habituation and anti-poaching project, and location of the film *Gorillas in the Mist*. A most memorable location it is too, protecting the upper slopes of the Virungas, a chain of free-standing volcanic

Mountain gorillas are the main attraction at Volcanoes National Park (AVZ)

Above A canopy walkway in Nyungwe, which protects the largest extant stand of rainforest in East Africa. (AVZ)
Left A chilly stream runs through the forest interior of Nyungwe National Park. (AVZ)

mountains whose daunting slopes, swathed in giant bamboo clumps and montane forest, rise imperiously to above 4,000m. Gorilla tracking, arguably the most emotive and haunting wildlife experience in the world, is the main activity here, with a total of 64 permits issued daily, and photographic opportunities are generally better than in Uganda, with the gorillas often hanging out in the open bamboo zone rather than the murky forest depths. Other guided activities include a day hike to Dian Fossey's abandoned camp and tomb on the forested slopes of Karisimbi, a visit to a habituated troop of the endemic golden monkey, and day or overnight ascents of the three tallest volcanoes.

Extending over 1,000km², **Nyungwe National Park** is the largest extant montane forest in East Africa and possibly the most scenic, covering a succession of tree-swathed mountains as they run south to the Burundi border. Accessible from the surfaced main road to Cyangugu, Nyungwe has few peers in terms of primate biodiversity, with troops of up to 400 Ruwenzori colobus easily located near the park headquarters, and a dozen other species present including L'Hoest's, red-tailed, Dent's and silver monkey, and an estimated 500 chimpanzees. Prominent among 275 bird species are the delightful great blue turaco and 24 Albertine Rift endemics, while a remarkable floral diversity includes 100 orchid species.

Rwanda's only savannah reserve is **Akagera National Park**, which protects the undulating woodlands and sprawling wetlands associated with the eponymous river on the Tanzanian border. Greatly reduced in size post-genocide to accommodate returned refugees, Akagera was also subjected to large-scale poaching in the 1990s, leading to the extinction of lion and rhino. The park remained in bad shape until 2010, when private management was brought in to fence it, curb poaching, improve roads, and generally nudge it back towards viability. Following an extensive programme of reintroductions, it is now home to all the Big Five (including both species of rhino), while boat trips are excellent for hippo and waterbirds.

Gazetted in 2015, **Gishwati Mukura National Park** protects two small patches of indigenous forest that harbour chimpanzee, golden monkey, L'Hoest's monkey and a varied selection of forest birds. Guided walks operate out of a community-owned guesthouse.

Plains zebra, Akagera National Park. (AVZ)

FURTHER READING

FIELD GUIDES

Several excellent field guides now cover Africa as a whole or East Africa specifically. The following titles are particularly recommended and are generally available from good bookshops or online.

MAMMALS

Pocket Guide to Mammals of East Africa. Chris and Mathilde Stuart. Struik (2011). A good option for space-conscious travellers who are serious about naming all the large mammals they see but can skip the smaller stuff.

Kingdon's Field Guide to African Mammals. Jonathan Kingdon. Bloomsbury (2nd Edition 2015). This is truly superlative: everything you need in a field guide and supplemented by detailed information about evolutionary relationships between modern species, all in a remarkably compact form.

Kingdon's Pocket Guide to African Mammals. Jonathan Kingdon. Christopher Helm (2nd edition, 2020). This compact and inexpensive title will meet the needs of most one-off visitors.

Field Guide to the Tracks & Signs of Southern, Central & East African Wildlife. Chris and Mathilde Stuart. Struik (3rd edition, 2019). Excellent for foot safaris, this will help you identify most of the tracks and scats you see along the way.

The Safari Companion: Guide to Watching African Mammals. Richard Estes. Chelsea Green Publishers (2nd edition, 1999). More a guide to mammalian behaviour than a conventional field guide, this superb book is well organised and informative, and aimed more at regular safari-goers than one-off visitors.

BIRDS

Birds of Africa South of the Sahara. Ian Sinclair and Peter Ryan. Struik (2nd edition, 2011). This ambitious title covers all 2,100-plus species recorded in sub-Saharan Africa in a remarkably compact 700-odd pages, with accurate illustrations, authoritative text and good maps. Not ideal for those whose travels are restricted to East Africa, but perfect for regular trans-African wanderers.

Birds of Kenya and Northern Tanzania. Zimmerman, Turner, Pearson, Willet and Pratt. Christopher Helm (2005). The first choice for any serious birder sticking to Kenya and northeastern Tanzania (it provides complete coverage of the northern Tanzania safari circuit, Usambara and Pare mountains and Pemba Island), this is a contender for the best single-volume field guide available to any African country or region. The gaps in its coverage limit its usefulness in Rwanda, Uganda and other parts of Tanzania. An older and more detailed hard cover version, now out of print, should be sought out by the dedicated.

Field Guide to the Birds of East Africa. Terry Stevenson and John Fanshawe. Christopher Helm (2nd edition, 2020). This excellent field guide provides comprehensive coverage for the whole of Kenya, Tanzania, Uganda, Rwanda and Burundi, with accurate plates, good distribution maps, and concise descriptions. The one birding 'must have' for Uganda, Rwanda, and any part of Tanzania west of the Serengeti or south of the Usambara. It is available as an app, too.

OTHER FIELD GUIDES

Amphibians of East Africa. Alan Channing and Kim Howell. Comstock Books (2006). An East Africa frogger's dream come true, this splendid newly published hardback is too bulky for casual safari use.

Field Guide to Common Trees and Shrubs of East Africa. Najma Dharani. Struik (2002). Although far from comprehensive, this is a useful field guide for those who want to put a name to the region's commoner trees.

Pocket Guide to the Reptiles and Amphibians of East Africa. Stephen Spawls et al. Christopher Helm (2006). This portable and authoritative guide covers 150 of the more common snakes, lizards, frogs, turtles and other scaly vertebrates likely to be seen on safari in East Africa. The author supplied some of the reptile pictures for this book.

BRADT GUIDES

Bradt publishes dedicated travel guides to each of the four countries in East Africa, combining comprehensive background information to individual game reserves and other sites of interest with all the practical detail you need to plan a successful trip. For details, see **w** bradtguides.com/shop.

Northern Tanzania Safari Guide: Philip Briggs and Chris McIntyre (5th edition, 2023)
Rwanda: Philip Briggs (8th edition, 2023)
Tanzania Safari Guide: Philip Briggs and Chris McIntyre (9th edition, 2023)
Uganda: Philip Briggs (10th edition, 2024)

COFFEE TABLE AND PICTORIAL BOOKS

Africa: Continent of Contrasts. Briggs, Harvey and Van Zandbergen. Struik (2005)
An African Love Story: Love, Life and Elephants. Daphne Sheldrick, Penguin (2013)
Big Cat Diary: Cheetah. Jonathan and Angela Scott. Collins (2006)
Big Cat Diary: Leopard. Jonathan and Angela Scott. Collins (2003)
Big Cat Diary: Lion. Jonathan and Angela Scott. Collins (2006)
Cats of Africa. Luke Hunter and Gerald Hinde. Struik (2013)
The Circle of Life: Wildlife on the African Savannah. Anup Shah. Harry N Abrams (2003)
Dotted Plains, Spotted Game: Images from the Masai Mara. Paul Goldstein and Roger Hooper. Crowley Esmonde Ltd (2005)
Great Ape Odyssey. Karl Ammann. Harry N Abrams (2005)
The Leopard's Tale: featuring Half-Tarl and Zawadi. Jonathan and Angela Scott. Bradt (2013)
Origin Africa: A Natural History. Jonathan Kingdon. Princeton University Press (2023)
Serengeti: Natural Order on the African Plain. Mitsuaki Iwago. Chronicle (1996)

APPENDIX ·

INTRODUCTION TO TAXONOMY

Taxonomy... is often undervalued as a glorified form of filing... but it is a fundamental and dynamic science, dedicated to exploring the causes of relationships and similarities among organisms. Classifications are theories about the basis of natural order, not dull catalogues compiled only to avoid chaos.

Stephen Jay Gould

Taxonomy is the branch of biology concerned with the classification of living organisms and the hierarchical representation of their relationships. Higher tiers of this hierarchy are known as kingdoms, phyla, subphyla and classes. All vertebrates, for instance, belong to the animal kingdom, phylum Chordata and subphylum Vertebrata.

Five vertebrate classes are conventionally recognised: mammals, birds, reptiles, amphibians and fish. Each of these is divided into a number of orders, which are further divided into families and then genera (singular genus), with intermediate ranks such as suborders, tribes and subfamilies being applied to more complex groups. All baboons, for instance, are placed in the order Primates, suborder Catarrhini (monkeys and apes), family Cercopithecoidea (Old World monkeys), subfamily Cercopithecidae (cheek-pouch monkeys) and genus *Papio* (baboons).

The Linnaean scheme of nomenclature, devised in the 18th century by the Swedish botanist Carolus Linnaeus, assigns every living organism a scientific binomial, which is a two-part name that indicates both its genus and species. Thus *Papio cynocephalus* is the yellow baboon and *Papio anubis* the olive baboon. These names are mainly derived from Greek, Latin or a combination of the two, and at species and generic level – though not above – they are always written in italics.

Linnaean taxonomy pre-dated Darwin's theory of evolution by a century, which meant that initial classifications were based largely on structural similarities between organisms. As evolutionary theory has gained greater acceptance, however, so the Linnaean scheme has undergone numerous modifications to reflect probable genetic and evolutionary relationships. An early victim of this paradigm shift was the 'thick-skinned' order Pachydermata (elephants, rhinos and hippos), which was widely recognised in the 19th century but has since been discarded for lacking any scientific basis.

Most modern taxonomists strive towards a scheme that is fully cladistic. In simple terms this means that every taxonomic designation should represent an evolutionary 'clade', a term used to describe a collection of living or dead organisms that descend from one common nodal ancestor. Biologists cannot always determine such evolutionary relationships with absolute certainty, but molecular testing, combined with more conventional anatomic, distributional and behavioural factors, makes it increasingly possible to theorise with reasonable accuracy. Mammals, for instance, almost certainly represent a true clade, as do the order Primates and the family Hominidae (apes and humans), since all these taxonomic groups comprise an ancestral form and every one of its descendants – though this would no longer be the case were one to follow creationist doctrine and remove humans from any of these groupings.

Taxonomy is a tool that aims to help us understand the natural order of things. Its limitations are shown up when one considers the process of speciation, which might occur over thousands of generations and – like any gradual process – will probably lack for any absolute landmarks. In simple terms, speciation starts when a single population is split into two mutually exclusive breeding units, which might happen as a result of geographic isolation or habitat differences. Whatever the reason, the two breeding units will share an identical gene pool when first they split, but each will accumulate a number of small genetic differences over generations, leading to clearly different characteristics that eventually mark them out as separate subspecies. Given long enough, the two populations will deviate to the point where they wouldn't or couldn't interbreed, even if the barrier that originally divided them is removed, at which point they would be classified as separate species.

This gradual process creates grey areas that no arbitrary distinction can cover: at any given moment there will exist separate breeding populations of a species that have not yet evolved distinct subspecific characters, and distinct subspecies that are on their way to becoming full species. Furthermore, where no conclusive evidence exists, some taxonomists tend be habitual 'lumpers' and others eager 'splitters' – respectively inclined to designate any controversial taxa either subspecific or full specific status.

For this reason, field guides often differ in their designation of controversial taxa. Our earlier baboon example is a case in point: olive and yellow baboons are regarded by some taxonomists as a single species (conspecific) on the basis that they often interbreed where their ranges overlap. These taxonomists thus assign them trinomials, *Papio cynocephalus cynocephalus* and *Papio cynocephalus anubis* respectively, in which the third part of each name indicates the subspecies.

Olive baboon: species or subspecies? (AVZ)

Such ambiguities can be a source of genuine frustration to the layman, and are particularly reviled by birdwatchers obsessed with ticking 'new' species. But they also serve as a valid reminder that the natural world is and will always be a more complex, mysterious and dynamic entity than any taxonomic construct designed to label it.

INDEX

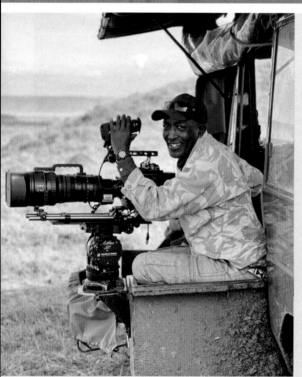